Praise for *Building Your Permaculture Property*

A book to take the reader from thinking into action, *Building Your Property* offers an excellent addition to permaculture theory and provides a key resource for all designers. By confronting and working through real, thorny, and often invisible human and landscape problems—a terrain in which they have earned their share of cuts and bruises—the authors slice through the Gordian knot that stops most people from realizing their home and nature visions through a powerful design system. Their vibrant and positive attitude yoked to psychological insight harnesses clean language and a keen focus on process to cut a neat furrow of systematic thinking through the complexity of living systems assessment, design, and management. Offering a window on digital design tools, clever illustrations, and examples from the demanding world of cold prairie farming, the authors have created a well-marked pathway for the advanced learner to reach professional outcomes.

— Peter Bane, executive director, Permaculture Institute of North America, author, *The Permaculture Handbook*

As well as being a valuable tool for individuals in their permaculture journeys, *Building Your Permaculture Property* represents another step towards permaculture being recognized beyond its widespread misconception of being simply a fashionable form of organic gardening. It highlights the need for permaculture design thinking in creating resilient, regenerative, landscapes and communities. Through this work, Rob, Michelle, and Takota make a valuable contribution to the ongoing evolution of permaculture thinking and action.

— David Holmgren, permaculture co-originator

A fresh, integrative, and holistic perspective on how to orientate oneself to the process of establishing your dreams and visions on the land. Designing and managing a farm that can build soil, create amazing food products, and sustain the farmer financially is possible anywhere; and yet it is the clarity of our context and decision-making and our attitudinal responses to design and management that largely underlie success. If you are dreaming of starting out on the land, this book will be a useful companion that will help you clarify your own approach to success, and help you navigate complexity with confidence.

— Richard Perkins, author, *Regenerative Agriculture*, owner, Ridgedale Farm AB and Making Small Farms Work AB

Exactly what is needed in the regenerative agriculture, homesteading, and permaculture community right now. It is loaded with practical, non-ideological, and actionable material on how to approach building your permaculture homestead. The information here is completely guided by experience and facts, unlike a lot of content out there these days that is rooted in hopes and dreams and built on a lack of practical application. This book will certainly be my manual as I develop my off-grid permaculture homestead.

— Curtis Stone, farmer, consultant, and author, *The Urban Farmer*

Those just embarking on a project, and those in the thick of creation, will find value from *Building Your Permaculture Property*'s systems approach to regenerative design. Anchored in permaculture's basic tenets—earth care, people care, and future care—the book's five-step process is organized with templates to collect and manage information, helping the reader to identify design ideas that consider her values, resources, and personal vision. These mental tasks are enhanced by playful illustrations by Jarett Sitter and lessons learned on the ground from the Coen family farm. Highly recommended!

— Catherine Wanek, co-founder, Builders Without Borders,
author and photographer, *The New Strawbale Home*
and *The Hybrid House*, and co-editor, *The Art of Natural Building*

As a well-being economist I would encourage government policy makers, First Nations, businesses, and farmers to contemplate the pragmatic processes and guidelines offered in this wonderful book. What would a permacultural approach look like adopted to the complex challenges of our economy? A permaculturist approach to the economy might consider maximizing well-being and minimizing suffering as the ultimate goal of a better life for all. This book provides a practical choreography of how to optimize well-being in our front yard gardens, our farmland, and across this vast expanse of Canada.

— Mark Anielski, economist and author, *The Economics of Happiness*
and *An Economy of Well-Being*

If you are serious about designing a permaculture property, this book has to be in your toolkit. The authors offer an accessible and current guide to the complexity of good design based on years of practical experience.

— Morag Gamble, Permaculture Education Institute

A critical book for those at the beginning of planning their permaculture property. Michelle, Takota, and Rob have spent years developing their five-step permaculture process and have now boiled it down into one concise manual. This book does an amazing job of breaking down and simplifying a complex design process and will get you on your way to building your dream permaculture property! Practical, concise, and an essential read!

— Jen Feigin, ex-director, The Endeavour Centre

A life well-lived includes leaving the land better than we found it. This five-step design manual jumpstarts that journey to a foregone conclusion, laying out a thoughtful process for making permaculture principles your own. Every farm, every ranch, and every homestead benefit from thinking deeper about how human intent engages with the places we're blessed to call home. Restoring integrity to degraded ground is our primal mission now as a species. Restoring diversity means planting many more trees. Restoring ecological posterity begins with listening to the heart of the mother…and then reading this book.

— Michael Phillips, Holistic Orchard Network, author, *The Holistic Orchard*

This book is doing a world of good by reorienting readers to the holistic nature of permaculture. More than just a gardening method, permaculture is rooted in systematic observation, design, practice, and feedback for re-design to create regenerative, biodiverse, and profitable landscapes. I enjoyed reading this book!

— Zach Loeks, director, The Ecosystem Solution Institute, author, *The Edible Ecosystem Solution* and *The Permaculture Market Garden*

Rob, Michelle, and Takota have put a pair of glasses on something that is often blurry in permaculture design: process. Their step-by-step process from beginning to end is exceptionally useful, along with Takota's story which proves the process through a case study of a well-functioning, finely-tuned permaculture farm. Interwoven with a good amount of philosophy and detail, *Building Your Permaculture Property* is a needed read for anyone who is serious about developing their property through a permaculture design.

— Nicholas Burtner, founder and director, The School of Permaculture

BUILDING YOUR
PERMACULTURE PROPERTY

[A FIVE-STEP PROCESS] TO DESIGN AND DEVELOP LAND

ROB AVIS, MICHELLE AVIS, TAKOTA COEN

new society
PUBLISHERS

Copyright © 2021 by Rob Avis, Michelle Avis, and Takota Coen.
All rights reserved.

Cover design by Diane McIntosh.
Cover photos © Takota Coen (main image) Insets: Rob Avis (excavator),
Gavin Young (seabuckthorn tree), Takota Coen (cows, farmyard)
Back cover photos © (top to bottom): Carolyn Fletcher, Takota Coen,
Takota Coen, Gavin Young, Gavin Young

Printed in Canada. First printing April 2021.

This book is intended to be educational and informative.
It is not intended to serve as a guide. The author and publisher disclaim
all responsibility for any liability, loss, or risk that may be associated
with the application of any of the contents of this book.

Inquiries regarding requests to reprint all or part of *Building Your Permaculture Property* should be addressed to New Society Publishers at the address below.
To order directly from the publishers, please call toll-free (North America)
1-800-567-6772, or order online at www.newsociety.com

Any other inquiries can be directed by mail to:

New Society Publishers
P.O. Box 189, Gabriola Island, BC V0R 1X0, Canada
(250) 247-9737

LIBRARY AND ARCHIVES CANADA CATALOGUING IN PUBLICATION

Title: Building your permaculture property : a five-step process
to design and develop land / Rob Avis, Michelle Avis, Takota Coen.

Other titles: Permaculture property

Names: Avis, Rob, 1979– author. | Avis, Michelle, 1980– author. |
Coen, Takota, 1992– author.

Description: Includes bibliographical references and index.

Identifiers: Canadiana (print) 2021012377X | Canadiana (ebook) 20210123788 |
ISBN 9780865719378 (softcover) | ISBN 9781550927306 (PDF) |
ISBN 9781771423267 (EPUB)

Subjects: LCSH: Permaculture—Planning. | LCSH: Permaculture.

Classification: LCC S494.5.P47 A95 2021 | DDC 631.5/8—dc23

Funded by the Government of Canada | Financé par le gouvernement du Canada

New Society Publishers' mission is to publish books that contribute in fundamental ways to building an ecologically sustainable and just society, and to do so with the least possible impact on the environment, in a manner that models this vision.

Contents

Foreword by Geoff Lawton xi

Preface xiii

Introduction 1
   The Problem with Permaculture 6
   You Need a Process (Not a Prescription) 11
   About This Book and the Companion Website 17
   Your Very First Practice: Get an Accountability Partner   19

Step 0: Inspect Your Paradigm 23
   The Gorilla in the Room 25
   The Upward and Downward Spirals 26
   Takota's Story: The Coen Permaculture Farm
      Upward Spiral 38
   Practices for Step 0. Inspect Your Paradigm 40

Step 1: Clarify Your Vision, Values, and Resources . . . 43
   What Do You Have? 45
   What Is Right? 50
   Takota's Story: Two Paths to the Same Cliff 54
   What Do You Want? 56
   Walking Through a Field of Landmines
      Blinded by a Scarf 60
   Be Careful What You Wish For 64
   Practices for Step 1: Clarify 67
   Takota's Story: Buckets of Well-being 68

Step 2: Diagnose Your Resources for Strengths,
   Weaknesses, Opportunities, and Threats 75
   Takota's Story: Don't Skip Your Diagnosis! 78
   A Watershed of Information 79

Takota's Story: Growing Up a Carpenter 80
Two Stages of Diagnosis 87
Black Swans 93
Takota's Story: Black Swan Dam 98
The Value of Digital Mapping and Open Data 99
Practices for Step 2: Diagnose 101

Step 3: Design Your Resources to Meet Your Vision and Values 105
Why Design? 108
What Design Is Not 110
Takota's Story: To Swale or Subsoil? 113
Form, Timing, Placement, and Scale 116
Takota's Story: The Good, the Bad, and the Ugly . . . 117
Creating a Permaculture Design 127
Practices for Step 3: Design 134

Step 4: Implement the Right Design That Will Most Improve Your Weakest Resource . . . 139
What Is Your Birdshot? 142
What Is Your Slug? 147
Pull the Trigger 151
Good, Bad, and Ugly Decisions 156
Takota's Story: The Bazooka Approach 157
Practices for Step 4: Implement 165
Takota's Story: My Best Advice for Solving
 Any Problem 166

Step 5: Monitor Your Resources for Indicators of Well-being or Suffering 169
The Push and Pull of Life 172
Monitoring Your Resources 175
Takota's Story: Monitoring for Mastitis 176
Takota's Story: Building My Own
 Permaculture Property 184
Takota's Story: An Ecosystem Disguised as a Farm . . 193
Practices for Step 5: Monitor 196

The Solution to a Sisyphean Task 199
    Putting It All Together 201
    Your Very Last Practice:
        Your Permaculture Property Planner 204

**Afterword: The Land Needs Us to
    Live Differently Here** 211

Glossary 215

Notes 217

Index 223

About the Authors 230

About New Society Publishers 232

Foreword

by Geoff Lawton

I had the pleasure of learning from and working directly with Bill Mollison, the co-founder of permaculture, and during our many years together, he passed along three pieces of advice that I offer to you now.

One morning in the mid-1980s, I was camped out in Bill's food forest. I hadn't slept all night because a question was bothering me. Finally, at dawn I climbed out of my tent and made my way toward Bill's home hoping to find an answer. The light was on in the window, and as I quietly approached, I could see that Bill was already at his desk ferociously reading and writing (he could actually do both at the same time). Although I hadn't intended to intrude, Bill spotted me quickly and beckoned for me to come in. "What's up?" he asked, once I had joined him inside. I replied, "I need advice… How do I know if I'm on the right path?" Bill contemplated for a moment, then answered: "You'll know if you are doing the right thing and on the right path if resources start to gather around you, and a lot of those resources will be really good people."

Bill was absolutely right. During my more than 30 years of working on the permaculture path, I have had some incredible people gather around me, and three of them are the authors of this book. I first had the pleasure of getting to know Rob Avis during his internship on Zaytuna Farm over a decade ago. On one particular night, I was awoken by the sounds of a very large tropical rainstorm thundering down onto the tin roof. I quickly started getting dressed as I knew that we were going to have big overflows from our dams, swales, and other earthworks, and I always like to keep an eye on them. As I was exiting the house onto the veranda, I was surprised to see a flashlight out in pouring rain coming toward me. "Who is that? What are you doing?" I called out. "Hi… It's Rob," replied a voice, "I just thought you might be going out." I chuckled and invited Rob to join me. So off we went to the top of the property, where we started trudging through the swales in thigh-high water. We were returning from having inspected several dam sites when suddenly Rob jumped up out of the water and yelled out, "What the heck was that?" I chuckled, and responded, "Likely an eel, mate!" Rob then burst into laughter and continued right along to check out the next spillway. I knew at that moment he was going to do great things!

Michelle and Takota are also past students of mine, which reminds me of something else Bill told me once: "If you are a good teacher, for every 350 to 400 students, you should have produced one permaculture teacher. And they, in turn, should go on producing permaculture teachers." At the time of writing, I have taught over 15,000 students, and I am proud to know that Rob, Michelle, and Takota are some of the most active educators in the permaculture space.

One of the last pieces of advice Bill gave me was: "You've gotta have dogged persistence." I in turn pass this on to all my students, and there is no question that this book is proof that Rob, Michelle, and Takota took it to heart.

I couldn't be more pleased to count the authors of *Building Your Permaculture Property* as past students, friends, and colleagues, and I'm very excited about this significant and unique contribution to the existing rich collection of literature on the subject of permaculture design. I'm confident that it will become a go-to resource for a significant number of permaculture designers and consultants.

— Geoff Lawton
January 2021

Preface

Many philosophers, academics, authors, colleagues, and friends helped us to tend the garden and care for the soil out of which this book eventually grew. Seeds were planted by many, starting initially with permaculture teachers Bill Mollison, David Holmgren, and Geoff Lawton. We also acknowledge other thought-leaders, Indigenous Peoples, and systems-thinkers who were planting, tending gardens, and sharing seeds and ideas many seasons before these teachers came along and inspired us. We've distilled many of these ideas collected from others and added a few ways of thinking that we've found particularly helpful in our own efforts to build permaculture properties. And although we've worked hard to make this a comprehensive and valuable resource, we'll be the first to admit that our book is biased, based only on our own experiences, and incomplete.

Building Your Permaculture Property is not an introduction to the concept of permaculture. If you have not taken a permaculture design course, have never picked up a book on permaculture design, or if you have not yet tried to put permaculture into practice on your own property, be sure to head to our website, mypermacultureproperty.com. There you'll find resources to help build your foundational knowledge so that you can get the most out of the information that you'll find here. And if you have no idea what permaculture is, then you definitely need to go and do some research elsewhere before diving in.

We also anticipate a fair number of readers to pick up this book and expect it to be chock-full of permaculture techniques. Let us resolve this now. In these pages you will find absolutely no mention on how to go about building a dam (or swales), installing food forests, choosing tree species, raising livestock, building home-scale rainwater harvesting, making compost, choosing renewable energy systems, or designing a low-energy home. Rather, *Building Your Permaculture Property* is about our Five-Step Permaculture Process,

Credit: Kurt Stenberg

a problem-solving methodology that includes step-by-step exercises, templates, productivity, planning and workflow tools, developed and honed through decades of successful land management consulting and farming. In many ways, this is the book we wished we had when we first started our own journeys.

For clarity and readability, we've opted to write the main text in the voice of Rob while most callouts are stories told by Takota from his own perspective. However, Takota, Michelle, and Rob have worked as a team to bring these ideas and this book to fruition, and we all consider ourselves to be equal contributors, authors, and partners in this project. All of the stories we tell about students and clients have been anonymized and changed as necessary to protect privacy. Further, to help ensure that everyone shares a similar understanding of key concepts, at the end of the book we have included a glossary of all words that have been bolded in the text.

While we are sincerely appreciative to our friends, reviewers, and colleagues who contributed and acted as sounding boards for this book, the biggest thank-you of all is owed to each of our parents, who built the foundation for each of us to be able to do this work. Lastly, we owe a huge thank-you to you, the reader, for your efforts in making the world a better place. We wish you all the best success, and hope that these ideas support you in building your ecologically regenerative, financially sustainable, enjoyable, and thriving permaculture property.

— Rob Avis, Michelle Avis, & Takota Coen
December 2020

Credit: Takota Coen

Our greatest hope is that our message, this book, will join the chorus of all those that have come before us and provide additional aid in our shared and ongoing attempt to learn to live differently on the land.

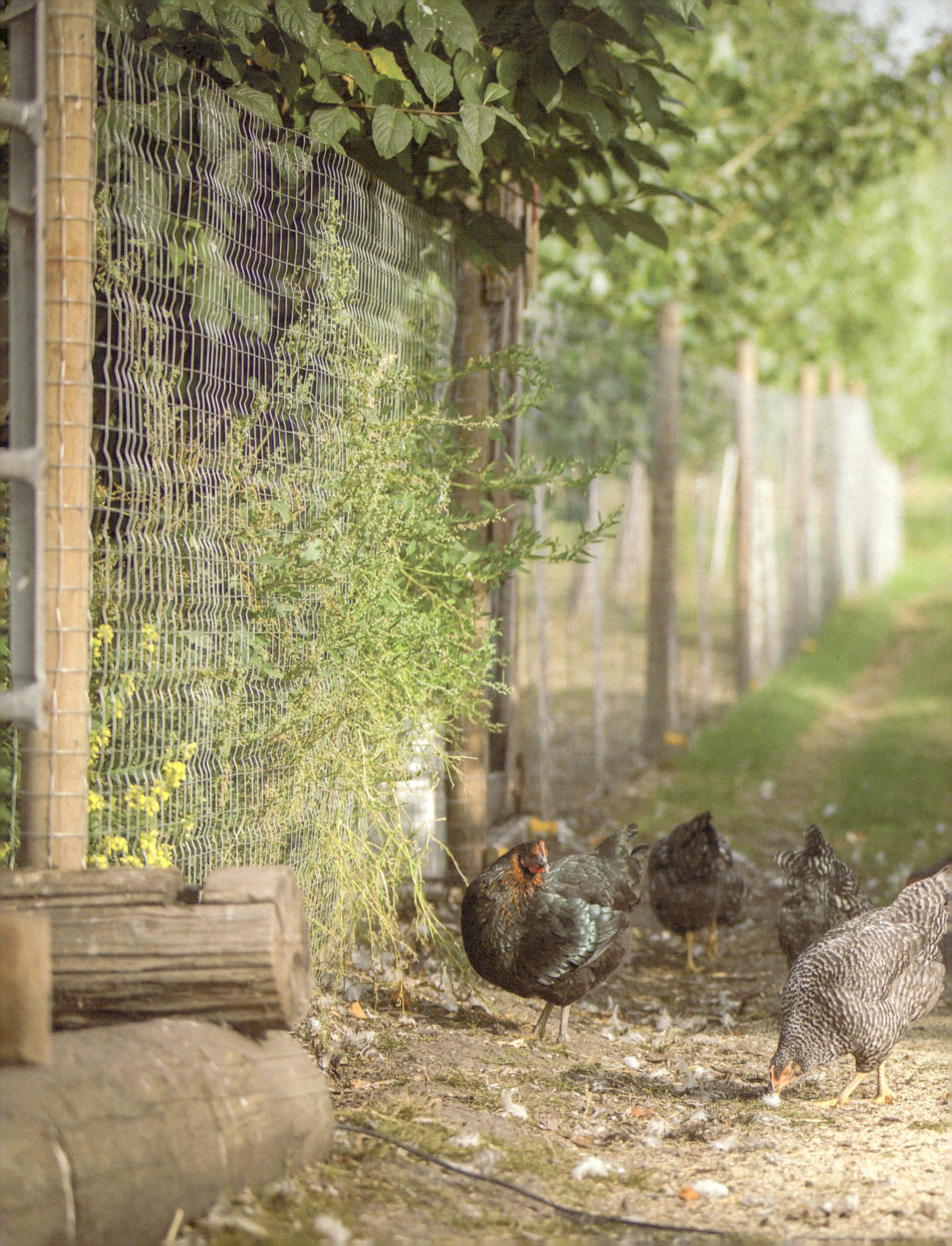

Introduction

Permaculture

per·ma·cul·ture

/ˈpərməˌkəlCHər/

noun

Permaculture is the conscious design and maintenance of agriculturally productive ecosystems which have the diversity, stability, and resilience of natural ecosystems. It is the harmonious integration of landscape and people providing their food, energy, shelter, and other material and nonmaterial needs in a sustainable way.[1]

Credit: Kurt Stenberg

I still clearly remember the day I decided to quit my job as an engineer in the oil and gas industry. I was at my computer and about to issue a request to have a multi-thousand-acre laneway of forest cleared in preparation for a new pipeline. With a flash of guilt and concern for the forest life, I hesitated issuing the request for a moment. On one hand, I disagreed with the destruction of this ecosystem, but on the other, I was an active consumer and petroleum products were the foundation of nearly everything I used and needed in my daily life. At that moment, an email from a friend arrived with a video link, providing the perfect distraction from my moral dilemma! I clicked through and started watching the five-minute YouTube video, *Greening the Desert*,[2] that would completely change the course of my life. It is the story about a man, Geoff Lawton, who had turned ten acres of desertified and degraded land in the Dead Sea valley into an oasis. I was amazed at what this man had accomplished, but confused by his final statement in the video: "You can fix all the world's problems in a garden."

Note that, at this point in my life, I had never grown a garden, or even been interested in gardening. The implication that all of the world's problems could be "solved in a garden" seemed overly simplistic, but there was something truly remarkable in that statement. Over the next few months, I became obsessed with understanding this metaphor and the idea of converting degraded landscapes into abundant ecosystems. I discovered that Geoff Lawton taught something called permaculture, and despite having invested four years of my life into studying engineering, the very definition of which is "the discovery, development or utilization of matter, materials or energy for the use and convenience of humans,"[3] permaculture design filled a gaping hole. In addition to actually providing methods and principles for the design of human habitat through an ethical framework, it provided a profoundly positive key insight. In Geoff's metaphor, "gardening" meant engaging with our surroundings in a way that provided for us, yet enhanced and improved ecosystems at the same time. In short, we humans could be as regenerative as we are destructive. I wanted to be part of this revolution.

A few years later, I found myself in a classroom on Geoff's farm in Australia on the first day of a permaculture design course. He walked in, wrote the words "Evidence to Act" on the board, and started an exposé about how fragile the world really was. I knew it

Introduction 3

Credit: Jarett Sitter

was bad, but as he described the global and accelerating trends of soil erosion, biodiversity collapse, ocean dead zones, food nutrient loss, peak oil, climate change, peak phosphorus, and more, I started to picture our entire civilization trapped between an immense cliff and a giant row of dominos streaming off into the distance. Any one of these issues seemed significant enough to cause a cascading failure, and in fact, in my daydream I could already hear the ominous sound of tumbling far off in the distance.

The rest of that course and time spent on Geoff's farm led me through cycles of grief: denial of how bad things really were, anger at all those "other people" who either created the problem or weren't

FIGURE 1. Perhaps you've also felt this same impending sense of doom: that our civilization is about to topple over the edge of a cliff.

FIGURE 2. (1) Avis family home in 2009. (2) Avis family home in 2015. (3) Michelle's and my other productive endeavour during this time. (4) My first passive solar greenhouse, heated with a rocket-mass heater.

> ### The Roots of Permaculture
>
> In the mid-1970s, Australian housemates Bill Mollison and David Holmgren started to develop ideas about stable agricultural systems. This was in response to the rapid growth of destructive industrial-agricultural methods. They saw that these methods were poisoning the land and water, reducing biodiversity, and removing billions of tons of topsoil from previously fertile landscapes. They announced their "permaculture" approach with the publication of *Permaculture One* in 1978.
>
> The term *permaculture* initially meant "permanent agriculture" but was quickly expanded to also stand for "permanent culture" as it was seen that social aspects were integral to a truly sustainable system (to find out about the roots of permaculture, get a copy of *Permaculture One*, Mollison and Holmgren, 1978). There have since been numerous other authors who have tackled the subject. Some of our favorite books include:
> - *Permaculture: A Designers' Manual*, Bill Mollison, 1988
> - *Introduction to Permaculture*, Bill Mollison, 1991
> - *Permaculture: Principles & Pathways Beyond Sustainability*, David Holmgren, 2003

doing something about it, bargaining for some way I could just escape to a bomb shelter in the woods, depression at the hopelessness of it all, and finally acceptance that if I was part of the problem, then I could be part of the solution! I returned home with hope and a new mission. I was going to take responsibility and get my house and garden in order to show a better way and lead by example.

Fast-forward a decade, and Michelle and I had transformed an inner-city home into a permaculture demonstration site complete with a passive solar greenhouse, renewable energy systems, micro-livestock, composting systems, a retrofitted hyper energy-efficient home, food forest, rainwater harvesting systems, and of course a productive vegetable garden. To date we have taught thousands of students through our education company, Verge Permaculture, published hundreds of blogs and videos, consulted on hundreds of projects, helped organize a permaculture community group, and most recently, moved and started developing a 65-hectare (160-acre) rural property.

Despite all this work, I must sheepishly admit that for years I was often ashamed to identify myself as a permaculture practitioner or to even mention the word "permaculture" in most professional circles because of a common misperception that permaculture is simply about growing a vegetable garden.

The Problem with Permaculture

When I was first introduced to permaculture design, I was blown away by the elegant, ingenious, practical, and seemingly simple solutions that Mollison and Holmgren proposed to solve the systemic problems for all of our food, water, shelter, and energy systems. But when I started doing more reading, visiting online forums and blogs, and taking courses, I encountered something quite different.

What I thought was a design system for all of our basic needs seemed more like example after example of gardening tips and tricks. And I know I am not alone in this because the most common response I get when I ask someone if they have ever heard of permaculture, if it isn't outright no, is ya, that's like organic gardening right? I shudder to recall how often I have heard certified permaculture teachers, designers, and even my own students, despite my best efforts, explain permaculture as "organic gardening on steroids."

While vegetable gardening is absolutely essential to permaculture, and I am not belittling its importance in the slightest, putting a herb spiral in your backyard while you still source the majority of your basic needs from the degenerative food, water, and energy systems is simply permaculture tourism. I liken this to claiming that you've "visited" Mexico, when in reality you simply spent two weeks in a gated five-star resort lying on the beach. What started out as a system to redesign every aspect of our human habitat based upon ecological and systems-thinking principles has the risk of transforming into nothing more than a collection of neat ideas for hipster gardeners looking for another hashtag to draw attention to their social media posts. In fact, if it weren't for the original teachings of Mollison, Holmgren, Lawton, and a few others I have had the privilege of studying under, I would have abandoned permaculture long ago. Sadly I know many former colleagues who have done just this. After years of struggling to put permaculture into practice, they throw their hands up and tell everyone "permaculture doesn't work." A statement that is, I think, as absurd as claiming "ecosystems don't work."

Why does permaculture get abandoned? Why has it not been more widely adopted by farmers, ranchers, builders, engineers, architects, and governments as an ethical solution to our growing global food, water, and energy problems? Why is it so often watered-down to "gardening on steroids"? These questions were keeping me up at night because their existence suggests that there was a problem with permaculture—something important was missing in how the teachings of Mollison, Holmgren, and others were being taught, understood, and practiced. I felt an urgent need to understand and address this problem, as I continued to see exponential growth in the ecological, social, and economic problems that I knew permaculture could solve.

The Biggest Struggles

With the above-mentioned questions in the back of my mind, I began to notice that whenever I was talking with a practicing permaculturist about their property, the conversation often focused on how challenging it was to put permaculture into practice! Hoping to get to the bottom of things, I began preemptively asking questions like *What is the biggest problem you are struggling with right now?* or *What is your weak link?* as often as I could. After hearing hundreds of answers, combined with requests for advice, it became evident that I could categorize the answers into five major themes. These themes provided my first clues to understanding the problem with permaculture, and for that reason, I call them the five struggles of permaculture.

Before I introduce these five struggles, let's do a little experiment. I want you to ask yourself the following question, ponder it, and then provide an answer, either out loud or in your mind before proceeding to the next section.

> *What is the biggest problem you are struggling with right now putting permaculture into practice?*

With a clear answer in your mind, lets see how it compares to the list in Table 1.

How did your own answer match up? Which biggest struggle best describes where you are at right now? Do you remember a time when you've experienced any of the other biggest struggles?

As you ponder these questions, consider the following patterns I have noticed about the five struggles of permaculture:

- People tend to move through all of the struggles sequentially, in the order I've presented, starting at the top row and moving downwards, although sometimes they skip over experiencing one or two of the biggest struggles.

How we garden reflects our worldview. When we see the world as a collection of independent and isolated elements, it is difficult, if not impossible, for us to grasp the interconnectedness of natural systems. How could we then garden ecologically, or live and act responsibly in an interdependent world?

Dave Jacke,[4] author and permaculture teacher

- If they manage to overcome the burnout in the fifth struggle, they start back at the top and cycle through again.
- It is normal for your answer to be a mixture of two biggest struggles if you happen to be in a transition phase. For instance, perhaps you are feeling slightly directionless (you don't know what you actually want) but also obsessed or excited (you don't have enough information) at the same time.
- If you feel your own answer is tied between rows that are not next to each other in the table and for instance you feel like you are both obsessed (you don't have enough information) and anxious (you don't know where to start), ask yourself which of the bordering biggest struggles you are closest to. If you are closer to

TABLE 1. The Five Struggles of Permaculture. The challenges and struggles you will encounter when building your permaculture property can be grouped into five categories.

Biggest Struggle	Common Questions Asked or Statements Made
I don't know what I should do.	‣ My partner(s) and I aren't on the same page. ‣ I can't figure out which of these elements I should choose. ‣ How big of a property should I get? ‣ Should I farm or homestead? ‣ Rural or urban? ‣ I'm not sure which direction to go. ‣ What would you do in my situation? ‣ I'm not sure what I actually want.
I don't know where to look.	‣ What is a good source of information about this topic? ‣ What are your favorite books / video channels / blogs / podcasts? ‣ What courses, conferences, or workshops should I register for? ‣ Which consultant should I hire?
I don't know how it all fits.	‣ How do I create a usable property design? ‣ How do I keep all this information straight? ‣ How do I know if my design is right? ‣ I'm worried that I've got it all wrong.
I don't know where to start or what's next.	‣ I'm not sure how to prioritize my time and money. ‣ I'm worried about not getting it right. ‣ My partner is really chomping at the bit. ‣ I don't want to paint myself in a corner.
I don't know when it will end.	‣ I feel like I'm spinning my wheels. ‣ I'm about ready to give up. ‣ Permaculture just doesn't work.

burnout (you are financially, physically, emotionally, or environmentally bankrupt) than being a little overwhelmed (you don't have enough information), then your biggest struggle is that you don't actually know what you want.
- If you feel like you could fit all of the categories, I press you to pick your biggest struggle. If you are still adamant that you're experiencing all five struggles right now, then you are most certainly in the burnout category, and you are reading the right book!
- If you feel confident that you have never experienced any of these struggles, then I guarantee that you have not been practicing permaculture long enough!

Although it is extremely helpful to categorize the biggest struggles this way, as you'll see soon, these struggles are really symptoms of something that is lacking. And that something is what this book aims to address.

The Six Ps of Epistemology

Before I can divulge what I believe to be the root cause of these struggles, I first need to introduce a model that describes any system of study that humans use to acquire and apply knowledge. I call this model **The Six Ps of Epistemology**:

Patterns. Repeating events that result from the interaction of two or more forms of energy or matter.

Paradigm. A set of unconscious assumptions that constitutes a way of viewing reality.

Philosophy. The systematic gathering of information about the interaction of energy and matter through the activity of logical reasoning.

Principle. A proposition that serves as the foundation for a system of belief or for a chain of reasoning.

Practices. The application of an idea, belief, or method.

Process. A series of steps taken in order to achieve a particular goal.

Each of these represents a different stage in which new information is discovered and eventually applied. The foundation of this model (as well as the source of all our knowledge) begins with the

---

epistemology

epis·te·mol·o·gy

\ i-ˌpi-stə-ˈmä-lə-jē \

The study of the nature of knowledge, especially with reference to its limits and validity.

occurrence of a *pattern* which is some kind of a repeating event that results from the interaction of two or more forms of energy, or matter, somewhere in the known universe. Electrons interact with protons, elements bond to each other, wind erodes rock, waves ripple sand, moons rotate around planets, planets rotate around suns, more of the same forms a galaxy, many galaxies become a universe, etc. It is important to note this interaction of energy or matter occurs regardless of our perception of it. This is objective reality.

Philosophy is the attempt to study and, hopefully, understand those infinite and objective patterns of the universe (the Greek word *philosophers* means "lovers of wisdom," they were the original scientists). However, the only way we can perceive the world is through

FIGURE 3. The interaction of two or more forms of energy or matter form patterns, which can be found everywhere you look throughout the universe.

Credit: Jarett Sitter

the subjective and ever-changing unconscious *paradigms* that are based upon our past experiences and our current mindset or beliefs.

From the subjective philosophy of the patterns of reality through the lens of our unconscious paradigms come conscious *principles* that serve as rules of thumb about how we believe specific aspects of how our world works and how we should act.

Those principles then give birth to *practices* we can take into the world to achieve our individual and collective goals.

And finally, as those practices, or events, are repeated over decades, even centuries, a new pattern begins to emerge. The form of this pattern is a hierarchical and chronological ordering of the best practices that reliably streamlines those activities. This is a *process*.

Given that *Permaculture: A Designers' Manual*,[5] was only published in 1988, permaculture is still a relatively new paradigm for the study of the patterns that make up our reality. While the principles upon which it is based come from ecosystems that are as old as the planet itself, the few decades permaculture has existed is not a lot of time for the infinite number of practices to be filtered down through trial and error into a step-by-step process that works reliably anywhere in the world. As such, much of the permaculture content available today is too focused on specific practices. The solution to all of the five struggles is a clear step-by-step *process*.

Much of the permaculture content available today is too focused on specific practices.

You Need a Process (Not a Prescription)

Many excellent step-by-step process-oriented resources have been developed that deal with individual elements of a permaculture system like forest gardening, annual gardening, natural building, water harvesting, and passive solar greenhouses. Some even describe step-by-step activities for designing a property on paper but fail to offer a plan for actually implementing that design or managing the property long term. Other resources deal with the process to manage a property holistically but have no mention of ecological design. To my knowledge, there has never been a complete and clear step-by-step process described for how to design, develop, and manage a property that provides for all your food, water, shelter, and energy needs in harmony with your surrounding ecosystem.

Think back again to your own permaculture reading, research, or training and notice how most books, courses, consultants, and content focus almost exclusively on one or more of the patterns, philosophy, paradigms, principles, and practices, *particularly the practices*, in the epistemological model but fall short when delineating a process.

There are even a few resources available where a permaculture practitioner develops a process that works for their property and then encourages others to copy what worked for their context. However, this step-by-step *prescription* often has unintended side effects!

It is no wonder that I and so many others struggled to put permaculture into practice, why few professionals and governments have adopted it, and why so many people say permaculture just doesn't work. I can empathize as I myself have made the mistake of blindly following the prescription of more than one of a growing number of gurus before I figured out that the root cause of overwhelm was not a lack of information, tools, technique, skills, or prescriptions, it was a lack of a **clearly defined process to design, develop, and manage a permaculture property**.

Clarify, Diagnose, Design, Implement, Monitor

I first met my co-author Takota Coen in 2014. He grew up on an organic mixed farm in central Alberta and after completing his first permaculture design course a few years prior had already spearheaded some major changes on his family farm. He called and asked if I would mentor him to start his own permaculture education and consulting business. Impressed with his initiative and his knowledge, I agreed, and it wasn't long before he was teaching in Verge Permaculture classrooms and co-consulting on our projects. We discovered that the combination of his on-the-ground practical organic farming experience and my process-oriented engineering background allowed us to tackle complex, multidisciplinary projects with a very unique perspective.

We began working on another project as well—to distill, articulate, and improve upon the process I had been using in my consultancy practice. We had a vision to be able to share this process—the way to think about and ultimately solve problems—in a manner that others could follow and apply to their own permaculture property projects. Leveraging our combined foundational experience, we tested ideas and tools with clients and their real-life projects. As we filtered and ordered our own practices, the patterns began to emerge, and five distinct phases, or steps, became apparent:

> **I myself have made the mistake of blindly following the prescription of more than one of a growing number of gurus before I figured out that the root cause of overwhelm was not a lack of information, tools, technique, skills, or prescriptions, it was a lack of a *clearly defined process to design, develop, and manage a permaculture property*.**

The problem is the solution.

Bill Mollison[6]

Step 1: Clarify your vision, values, and resources.
Step 2: Diagnose your resources for strengths, weaknesses, opportunities, and threats.
Step 3: Design your resources to meet your vision and values.
Step 4: Implement the right design that will most improve your weakest resource.
Step 5: Monitor your resources for indicators of well-being or suffering.

Soon we began teaching this process and sharing the accompanying tools to small groups of students. To our surprise, our students started producing designs that exceeded anything we, as consultants, could have ever created for them, and best of all, they were successfully implementing those designs. When we finally stood back and asked ourselves why does it work, it dawned on us. Each of the five steps corresponds directly to the five most common struggles we had noticed for years in our clients and students, and through our own personal experiences practicing permaculture. Table 2 shows each step in relation to the biggest struggles it addresses.

As our students followed this process with great success, we not only realized that it addressed these five common struggles, we also had a second major epiphany. In our private consulting engagements, Takota and I both had to spend hundreds of hours to get acquainted with the land, identify the client goals, translate it all into a design, coordinate the implementation, and then finally train the landowners to manage it all for themselves! This was not only cost prohibitive for most land stewards, it was also hard for us to scale our land regeneration efforts. Distilling and teaching our own internal process to others meant we could help an exponentially larger number of people, and motivated students were producing better designs than we could have ever done for them, for a fraction of the cost. It is for this reason that we say the best person to design and manage your land is you. No one knows you (and what you want) better than you do. No one knows your land better than you do. And no one else is going to be interacting with your property, your own masterpiece, in an ongoing way like you will be.

So, how then, do you actually use this process to create your own masterpiece?

TABLE 2. The 5SP Process. Each step in our Five-Step Permaculture Process (5SP Process) addresses one of the five biggest struggles with permaculture.

The 5SP Process	Corresponding Biggest Struggle
Step 1: Clarify your vision, values, and resources.	I don't know what I should do.
Step 2: Diagnose your resources for strengths, weaknesses, opportunities, and threats.	I don't know where to look.
Step 3: Design your resources to meet your vision and values.	I don't know how it all fits.
Step 4: Implement the right design that will most improve your weakest resource.	I don't know where to start or what's next.
Step 5: Monitor your resources for indicators of well-being or suffering.	I don't know when it will end.

Mimicking Michelangelo

When the Renaissance artist Michelangelo Buonarroti was asked how he was able to create his immortal 5-meter (17-foot) tall statue of *David*, he is reported to have replied, "It was easy; all I did was chip away the stone that didn't look like David. I saw the angel in the marble, and I carved until I set him free."[7]

While I know nothing about carving marble, this quote beautifully captures the spontaneous, intuitive, common sense, even mystical experience of designing and developing a permaculture property as more of a process of elimination that seeks to uncover what was already there than a process of creating something new. But this quote and many others about the creative work all seem to leave out two important aspects.

The first aspect missed that is easy to forget is the dogged persistence required to complete a masterpiece of any kind. In Michelangelo's case, he had a love of working with marble from an early age and spent his first twenty-six years honing his artistic skill in his father's marble quarry, and subsequently at the best art schools of his day and by completing numerous famous commissioned works, before he ever laid hands on the rock destined to become *David*. It took Michelangelo two years to complete *David*, and various biographers reported that he labored constantly during that time, rarely eating, sleeping only sporadically with his clothes and boots still on, sometimes within sight of his statue. Even when it rained, he continued to work in the open courtyard. Both Takota and I can relate to this kind of obsession and endurance. To this day, I still wake in the middle of the night with an idea that demands I jump out of bed to either write it down, confirm its validity by looking at my designs, or going for a moonlit walk across my own emerging sculpture. Just as in the case of *David*, this commitment more than pays off. I have long since lost count of the value of the opportunities seen and problems averted by always sleeping at the foot of my statue.

The second, and most important, is that experts always have a process that took thousands of hours of painstaking trial and error to develop and master. In the case of Michelangelo, it is difficult to prove what his exact process was because he destroyed much of the evidence of how he worked, most likely to conceal his methods from rival artists. But some artifacts reveal a workflow that likely entailed first creating rough sketches on paper often using molds of human body parts to ensure anatomical correctness. Those sketches would then be turned into a small terracotta model presented to his patron for approval. Michelangelo would then create detailed drawings for the marble slab he needed and would often oversee the stonecutters in the quarry to ensure the quality and safe transport of raw materials. His sketches were transferred onto the rough marble using plumb bobs, measuring sticks, and calipers. He would then create a miniature three-dimensional model, out of wax. This, presumably, enabled him to warm and reshape his original design as

Experts always have a process that took thousands of hours of painstaking trial and error to develop and master.

implementation progressed so that any unforeseen problems could be adapted into unimagined possibilities. But it also allowed him to submerged the wax model in water so that, as the level was dropped incrementally, it was possible to use the waterline both as a filter to hide the unnecessary details and as a reference point for the next strike of hammer and chisel on the real sculpture.

These two aspects of persistence and process are also often missing from the creative nature of building a permaculture property. Fortunately, the Five-Step Permaculture Process (5SP Process) addresses both, as explored below:

- **Step 1: Clarify** is about making sure that you actually want to be a sculptor and, from there, what you'd like your sculpture to be.
- **Step 2: Diagnose** is all about gathering and organizing the tools and information you need to sculpt, as well as finding the perfect slab. It's also about putting in the work and sleeping at the feet of your statue to make sure you don't miss any insights it whispers to you about how best to set it free.
- **Step 3: Design** is about letting your blank slate influence you as much as you influence it, in effect seeing the angel of the design inside the rock of diagnosis. Then making mistakes on paper or, better yet, adaptive three-dimensional models instead of in stone or on the land.
- **Step 4: Implement** is about strategically chipping away at the emergent design one piece at a time until that angel trapped inside is finally liberated.
- **Step 5: Monitor** is about using suffering and well-being as your waterline to keep the work on track yet enable quick adaptations to your wax model as necessary.

The cumulative effects of persistence and process aided Michelangelo to build his *David*, and these two things apply just as much to building a permaculture property as they do to sculpture. If you're looking to move past the obstacles that have been blocking you from creating your ecologically regenerative, financially sustainable, functionally resilient, and enjoyable property, our 5SP Process may be just what you need. Let me now explain how this book and its companion website are designed to guide you in learning this process.

About This Book and the Companion Website

I've spent this introductory chapter describing why it is usually not a lack of information or technique that makes designing and managing land so difficult. What's missing is a clear step-by-step process.

The next chapter is called Step 0: Inspect Your Paradigm. Although not technically part of our five-step process, your beliefs are fundamental to the reality you create, and as such heavily influence your success with your permaculture property. This chapter is also a crucial discussion on how a global paradigm shift is likely the only action that will save us from our own societal collapse. Each of the subsequent five chapters describes one of the steps in the 5SP Process. Following the epistemology model presented earlier, each chapter will start with a discussion of the patterns, paradigms, philosophies, and principles underlying that step. In the last part of every chapter, we share some specific practices that we (or others) have developed and that we use personally, teach our students, and/or find particularly useful in our consulting engagements. We'll share questions, exercises, worksheets, workflow tools, and templates. The practices are all numbered, both because we encourage you to tackle them in sequential order and because it makes it easier for us to refer to them throughout the book, as well as on the website. Know, however, that it's not at all important for you to mimic our exact practices. If a particular exercise doesn't work for you, you are strongly encouraged to modify it or even develop your own that suits your individual needs. Just make sure that it meets the chain of reasoning (i.e., the principles) of that step. You will likely even realize that some things that you are already doing will fit nicely within the 5SP Process. I'll close out the book with a system to plan, prioritize your tasks, track timelines, and keep you organized by putting everything presented here together in your own *Building Your Permaculture Property Planner*.

While the linear nature of reading and learning necessitates that I present each of these steps one at a time in specific order, it is important to note that in reality there is often an iteration of all the steps simultaneously. Just like playing a guitar isn't reading music, then keeping time, then strumming a rhythm, then making a chord: you have to learn all of these things independently, but to make

> If you give up on trying to change larger structures and just go off on what some would say is a personal indulgence or being a survivalist, it can be seen as incredibly negative or pessimistic. But the other way to think of it is this: through manifesting the way we live and acting as if it's normal, you're defending yourself against depression and dysfunction, but you're also providing a model that others can copy. And that is absolutely about bringing large-scale change.
>
> David Holmgren,[8] co-originator of the permaculture concept

music, you have to put it all together all at once. More than that, this process is not just learning how to copy someone else's song, it's about learning how to write and play your own solo. And you'll have to read the book a few times over, and commit to the process and associated practices for at least a year, before you'll likely feel like you can conduct a symphony.

To help you get there faster, the companion website, mypermacultureproperty.com, is where you'll find troves of additional resources, downloadable templates, bonus materials, our latest musings, updates to this book (perhaps even updates to our process!), and more. If the Greek philosopher Heraclitus will continue to be right in his observation that "change is the only constant," then our only hope of developing a clear step-by-step process for surviving and thriving in a complex world is making sure that it can be adaptive to those changes. And while everything in this book has been tested in our personal lives and professionally with our students and clients, new patterns and new understandings of old patterns are constantly emerging. We will endeavour to keep our website up-to-date with these ongoing evolutions.

If you take nothing else away from this book, remember this: you do not have to know exactly what kind of food, water, shelter, and energy systems are the best for creating resilience and abundance for your permaculture property, never mind the entire planet. Concentrating on this would be like walking across the bridge in Figure 4, and focusing your attention on the destination that is a great distance off and guessing where your next step should be. I guarantee that using this approach to design a property, let alone to fix our entire global fragilities, will bring about a quicker collapse.

Instead of focusing only on the destination, you need to focus on how to best put one foot in front of the other towards the better future we all know is possible. You need a process because, like Canadian designer Bruce Mau[9] wrote:

Process is more important than outcome. When the outcome drives the process, we will only ever go to where we've already been. If process drives the outcome, we may not know where we're going, but we will know we want to be there.

Instead of focusing only on the destination, you need to focus on how to best put one foot in front of the other towards the better future we all know is possible.

Your Very First Practice: Get an Accountability Partner

Before you get started with any practice from any of the steps, I want you to do a simple but crucially important action: go find yourself an accountability partner.

The best accountability partner will be someone who is also building their own permaculture property and not directly involved in your project (i.e., not your fellow land steward, your partner, parent, child, etc.). You are looking for a mutually beneficial peer-to-peer relationship where you can set up regular check-ins, discuss your goals and the practices in this book, and hold each other accountable to getting things done. Not only will these conversations help you feel motivated and supported, they are also important for you to be able to have your own insights and make your own connections about what you are doing and learning.

There's real science behind this, and it has been shown that peer pressure is powerful, especially when we are making complex decisions. When we started including this practice in our workshops on permaculture land design, the resounding feedback from students was that having an accountability partner was one of *the best things* about the process. Although I can't easily pair you up with someone else who is currently reading this book, the internet has made it infinitely easier for you to find someone yourself. And with numerous free platforms for video calling and computer screen sharing, it has never been simpler to connect with someone—even if they live on the other side of the planet. So don't be shy. Head to your local permaculture gathering, an online forum, or a social networking site and let others know that you are designing your permaculture property and are looking for an accountability partner. You'll likely be surprised (and thrilled) with the interest and the response as many people already recognize the value of this practice. You can also check our website for resources and opportunities to connect with folks there, but whatever you do, don't skip this practice.

FIGURE 4. (*over*) If you take stock of your surroundings, you will notice three very hopeful facts: the terrifying cliff you are standing before is actually a canyon and the other side is filled with unimaginable possibilities of ever increasing well-being, the sturdy bridge of permaculture has been shuttling people across the gap for decades, and this book provides you with a process to avoid the five struggles of permaculture that pose as your biggest stumbling blocks to building your own permaculture property.

20 Building Your Permaculture Property

Credit: Jarett Sitter

Introduction 21

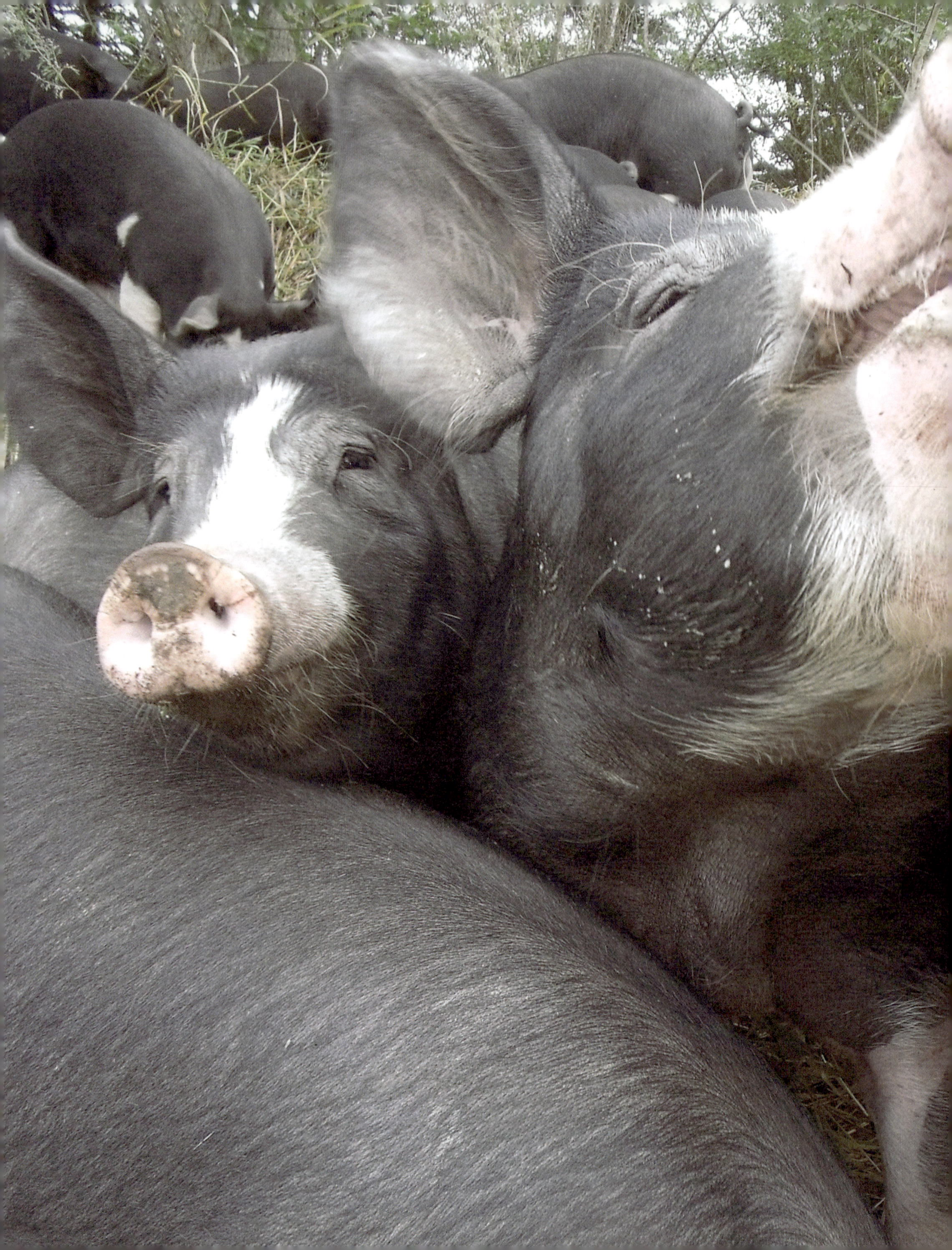

- How various cultural narratives have influenced the predominant conscious and unconscious paradigms, and the behavior, of our global civilization.
- Why we must evolve our collective paradigms to facilitate the successful co-creation with the complex systems of this world.
- Why I remain hopeful.

The Gorilla in the Room

There's an excellent two-minute video titled *The Monkey Business Illusion*[4] on YouTube that powerfully demonstrates how our paradigms are both a gorilla in the room that needs to be addressed and the very reason we can't see the gorilla. It begins with a black screen and single goal in text: "Count how many times the players in white pass the ball." Up pops six people on a stage, three of them wearing black shirts and three wearing white. One person from each color group has a ball. The players begin milling about and passing their ball to the players of the same color. It is difficult to keep track, but if you concentrate hard, after about thirty seconds, the players all leave the frame, and the answer to the goal at hand is provided: sixteen passes. Almost everyone gets it. But then another question is posed: "Did you see the gorilla?" About fifty percent of viewers will think this is a very strange question. But the reality is that partway through the game, a full-sized human, wearing a gorilla suit, walks clear through the group of players and pauses a few seconds to beat his chest. And even if you did spot the gorilla, you very likely didn't notice another scene change that occurs.

This experiment, and others like it, are designed to reveal patterns of error in our paradigms that are commonly referred to as cognitive biases. Not seeing the gorilla, and the other changes in the video, because your focus was elsewhere is an example of an attentional bias. So far scientists have identified dozens of these patterns of error such as the following:

- A confirmation bias, often deemed the most insidious of cognitive biases, is the tendency to search for, interpret, focus on, and remember information in a way that confirms one's preconceptions (i.e., paradigm). Can also be thought of as remembering the hits but ignoring the misses.[5]
- The endowment effect (e.g., loss aversion) is the tendency for people to demand much more to give up an object than they would be willing to pay to acquire it.

> Everything that we see is a shadow cast by that which we do not see.
>
> — Martin Luther King, Jr.[3]

> - The IKEA effect (my personal favorite) is the tendency for people to place a disproportionately high value on objects that they assembled themselves, such as furniture from IKEA, regardless of the quality of the end product.

It is very important to note the fact that we really do see the world through rose-colored glasses or some other frame of reference is not evidence that humans are stupid, lazy, careless, and immoral. The human brain receives 11 million bits per second of information, yet it appears to be only conscious of fifty bits per second,[6] or 0.0005% of its capacity. The function of our paradigms is therefore to conserve energy in our struggle to figure out how to live in the *present* moment. However, a paradigm becomes problematic when it no longer contributes to our well-being. Given that humans have evolved as a species much in part because of our ability to mentally time travel to events that are yet to happen,[7] we are undercutting our evolutionary traits by subscribing to the current paradigm, which is clearly not contributing to the long-term well-being of our species, and the others we co-inhabit this planet with. When we fail to be aware and consciously challenge our paradigms, we enter into the dangerous realm of dogma. And as the joke goes, "If we aren't careful, our dogma will get run over by our karma."

Have you ever noticed when wearing sun- or eyeglasses that it's possible to be completely unaware of smears on the lenses? And suddenly you notice the dirt, and then it's all you can see? In a similar fashion, it's possible to inspect our paradigms. And when we do this, we just might see how smudged, dirty, and cracked they really are, and opt to clean them up or swap for a new pair altogether in what is referred to as a paradigm shift.

Now that you have firsthand experience that "we can be blind to the obvious, and we are also blind to our blindness,"[8] let us take a closer look at some of the more common lenses through which humans see the world, how this influences our reality, and how we might go about shifting it. In other words, we can shape our paradigm before it shapes us.

The Upward and Downward Spirals

In *Collapse: How Societies Choose to Fail or Succeed*,[9] Pulitzer Prize-winning author Jared Diamond identifies the following five fundamental factors for how societies choose to fail or thrive:

1. Human-caused environmental damage
2. Planetary-scale climate change
3. Hostile neighbors
4. Loss of trading partners
5. Society's responses to its environmental problems

While the factors of planetary climate change, hostile neighbors, and the loss of trading partners all played a compounding role in some of the book's featured collapsed societies, it was human-caused environmental damage and the response to those problems that were the causes shared by all of the failed civilizations.

To know about the severity of our environmental damage today and to see the same lack of a meaningful response that brought the Greenland Norse, Easter Islanders, Pitcairn Islanders, Anasazi, and Maya civilizations to their knees begs some difficult questions. Are we as a species simply stupid, lazy, careless, and immoral? Are humans inherently bad? Despite the claim that you've likely heard that planet Earth would be better off without humans, I don't think this is the case. I believe that our inability to partner and thrive within the ecosystems on this planet has far more to do with destructive paradigms.

There are two predominant paradigms, which I call degenerative and sustainable, that are not only contributing to the struggle of permaculture property design but also keeping us on a similar trajectory of collapse, which I refer to as our **downward spiral**. There's also a third, *regenerative* paradigm that is a prerequisite to successfully building a permaculture property as well as starting the much-needed global-scale **upward spiral**.[10]

Before I can describe each of these paradigms and the upward/downward spirals in detail, we need to take a little sidestep and better define the starting point. Shown in the center of the following figure, that starting point is *complexity*.

The Complications of Complexity

First let me start with some definitions. A *system* at its most fundamental level is simply a bunch of interconnected and interdependent elements organized in a way to achieve some overarching function or purpose. Ecosystems are *complex systems* because they have the capacity to evolve. They are diverse, compounding, dynamic, emergent, and self-organizing. They can learn and create and

> You can hit a nail on the head, or cause a machine to do so, and get a fairly predictable result. Hit a dog on the head, and it will either dodge, bite back, or die, but it will never again react in the same way.
>
> Bill Mollison[11]

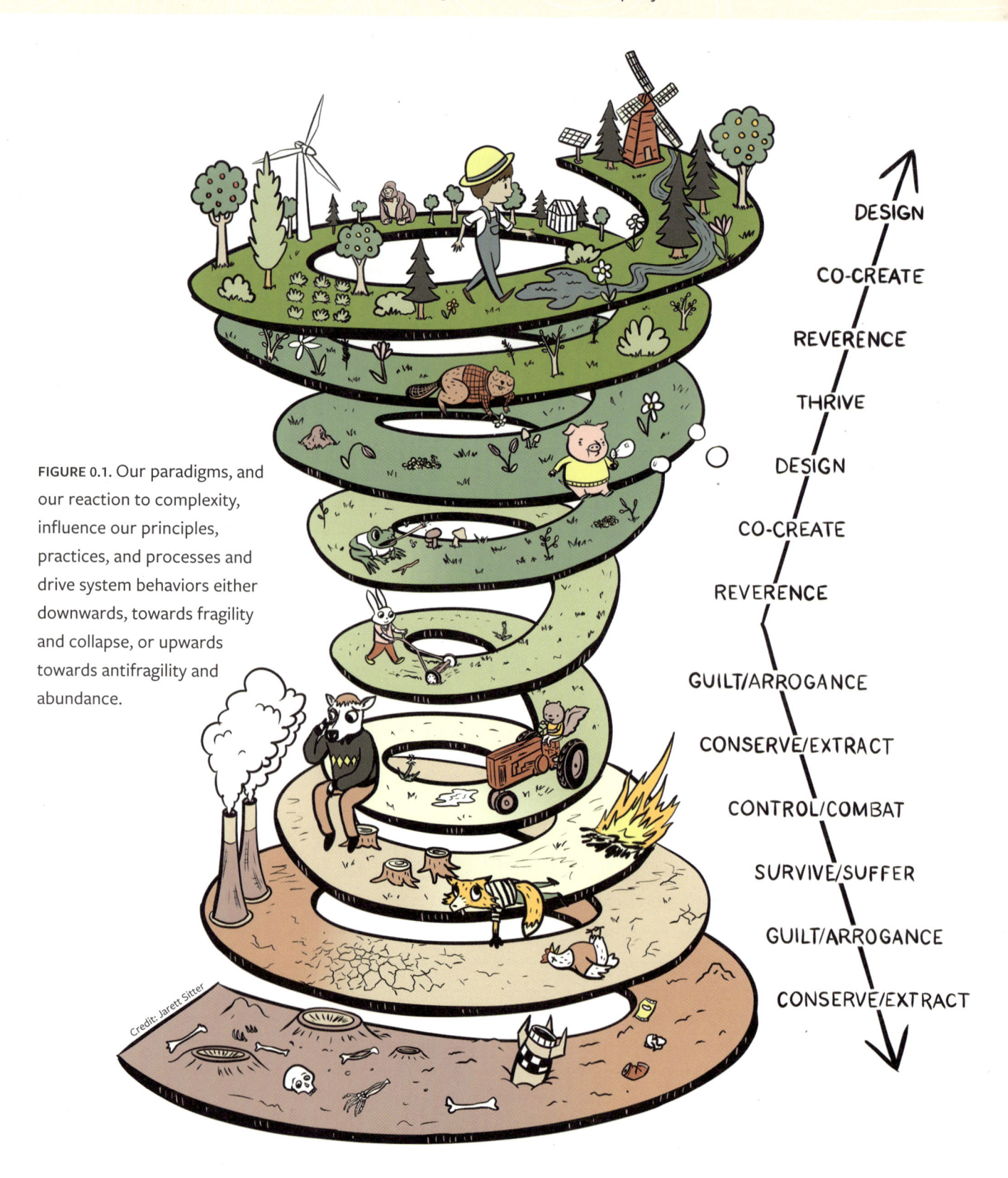

FIGURE 0.1. Our paradigms, and our reaction to complexity, influence our principles, practices, and processes and drive system behaviors either downwards, towards fragility and collapse, or upwards towards antifragility and abundance.

Credit: Jarett Sitter

DESIGN
CO-CREATE
REVERENCE
THRIVE
DESIGN
CO-CREATE
REVERENCE
GUILT/ARROGANCE
CONSERVE/EXTRACT
CONTROL/COMBAT
SURVIVE/SUFFER
GUILT/ARROGANCE
CONSERVE/EXTRACT

often have nonlinear and unpredictable behaviors. All biological and social systems are complex systems, a tree and the forest it calls home are complex systems, all of which contain many smaller complex subsystems. Perhaps you've already guessed where I'm going with this: your permaculture property is indeed a complex system. It contains material resources such as your house and any other physical infrastructure interconnected with many living resources, such as gardens, a food forest, animals, and many other biological elements, including you and your family members or partners involved in the project. *Permaculture*, with the appreciation of systems now in mind, can be defined as the deliberate design of complex systems, to meet all our human needs, with the overarching goal of benefiting all forms of life on planet Earth. And this is why I'm taking the time to make sure you understand what a complex system is, and subsequently how systems work!

Systems have another very important characteristic, described by an ability, or lack of ability, to withstand volatility or stress. I like to use the author Nassim Nicholas Taleb's concepts of fragile, resilient, and antifragile to describe a progressive scale of ability to withstand volatility.[12] Fragile systems break, resilient systems resist breaking, and antifragile systems actually get better, or benefit from some dose-dependent amount of stress. Take a moment and consider a centralized water supply system, drawing its source from a prairie-fed river. Would this system be able to supply water to thousands of people in the event of a multiyear drought? What about in a one-hundred-year flood? Likely not. This is a fragile system, and any household within this system with no secondary water supply is also fragile. By adding rainwater harvesting or perhaps a groundwater well with backup power supply, the homestead could be perhaps considered resilient. If a majority of residents did the same, then the municipality water supply might also be resilient. A property owner interested in creating an antifragile water supply would, in addition to securing backup water, engage in beaver restoration as well as groundwater recharge efforts, erosion control, and water-harvesting earthworks on her own property as well as supporting efforts throughout the watershed. She would intentionally design for increased life and increased complexity, knowing that living systems, through their inherent process of self-organization, continually increase the overall antifragility of the system. That way, if a

> *Permaculture*, with the appreciation of systems now in mind, can be defined as the deliberate design of complex systems, to meet all our human needs, with the overarching goal of benefiting all forms of life on planet Earth.

flood came, a greater amount of water would be infiltrated, and in a drought, the beavers would build more dams to slow what little water there was, better preparing themselves and the system for the next flood. Donella Meadows,[13] co-author of the 1972 best-selling publication, *The Limits to Growth*, says, "When you understand the power of system self-organization you begin to understand why biologists worship biodiversity even more than economists worship technology." I love to requote Meadows but replace the word "biologist" with "permaculturalist."

With all of these concepts and terminology out of the way, I can now bring it all back to the paradigms and the spirals that I introduced at the beginning of this section. How someone reacts or responds to the evolving, emergent, unpredictable, and self-organizing capacity of complex systems (i.e., complexity) will change greatly depending on their paradigm. The spirals shown in Figure 0.1 are repeating cycles of reactions and responses. The degenerative paradigm cycle is *arrogance—extract—combat—suffer*, spiralling downwards towards increased fragility and system collapse. The sustainable paradigm cycle is *guilt—conserve—control—survive*, still spiralling downwards, although slower, also towards increased fragility and collapse. The regenerative paradigm cycle is *reverence—co-create—design—thrive*. Only this last pattern of reaction and response to complexity leads "upwards" towards antifragility and abundance on your permaculture property.

The Degenerative, Sustainable, and Regenerative Paradigms

Table 0.1 expands on the degenerative, sustainable, and regenerative paradigm beliefs, visions, values, principles, practices, processes, and subsequently the trajectory of the complex system(s) being influenced.

The degenerative paradigm is based in a belief that humans are separate and better than nature, gods or godlike in a universe that exists for the sole purpose of benefiting humans. This superiority complex, arrogance, and fear of scarcity leads to unrestrained resource extraction and a win-lose game orientation: *He who dies with the most toys wins?* or *How do I get more goods?* Any destruction, chaos, and suffering to ecosystems and organisms that result is an acceptable externality. Equally disturbing is that harm inflicted on other humans can be ignored so long as they are not in the same

> **Black and White Versus Shades of Gray**
>
> I'll acknowledge that I describe the degenerative, sustainable, and regenerative worldviews as black or white, this or that, here or there, those people in this paradigm, these people in that paradigm. Although many humans may fall firmly in one of these categorizations, reality is far more nuanced than this simplification of ideas. However, even if you don't consider yourself squarely in one paradigm or another, watch out for undertones, influences, ideas, and dogma that foster these ways of thinking and that may be unconsciously shaping your worldview.

class, race, religion, or lifetime. The degenerative paradigm modus operandi when it comes to interacting with systems is to dominate and combat complexity.

In response to the self-righteousness of the degenerative paradigm, a polar opposite belief underlies the sustainable paradigm. While humans continue to be separate from the natural world, this time they are inferior to it. The self-loathing implied by this belief is perfectly captured by Sir David Attenborough when he stated that "humans are a plague on the earth."[14] This paradigm is founded on a principle of sufficiency and an unshakable faith that there could be just enough resources for everyone if humans would accept their wickedness, repent, restrain, and finally become virtuous by the masochistic practice of conservation driven by the question, *How*

TABLE 0.1. The Three Paradigms. The degenerative, sustainable, and regenerative paradigm beliefs, principles, practices, processes, and subsequently the trajectory of the complex system(s) they influence.

Paradigm	Principles based on:	Practices and processes based on:	Outcome/Trajectory:
Degenerative (Better than nature)	Arrogance, self-righteousness, dominance, scarcity, separation from nature, denial or willful ignorance, instant gratification, unimpeded exponential growth, dogma, fear	Sadistic extraction Combatting complexity	Suffering Fragility Collapse
Sustainable (Worse than nature)	Guilt, self-loathing, sufficiency, separation from nature, powerlessness, repentance, restraint, fear	Masochistic conservation Controlling complexity	Surviving Resilience Delayed collapse
Regenerative (We are nature)	Reverence, awe, abundance, co-operation, justice, humility	Compassionate co-creation Design for complexity	Thriving Antifragility Abundance

do I do less bad? Sustainable paradigm subscribers believe that the best way to engage in complex systems is to try and conserve and sustain things the way they are, subsequently they invest much time and effort attempting to suppress and fight against complexity and self-organization.

In the regenerative paradigm, humans see themselves not as separate and superior to nature, nor as separate and inferior to nature, but simply as members of a community of sentient life-forms that are all working towards the collective and interdependent goal of well-being. Under the right care and design, and by continually asking *How do I do more good?*, holders of this worldview believe that resources can be as abundant and infinitely expanding as the universe itself. Reverence and awe underpin practices based on compassionate win-win relationships with all living things. Righteousness is replaced with justice, dualism is replaced with oneness, and arrogance is replaced with humility. The logical intervention strategy for these self-described bio-collaborators is to intentionally design for increased complexity.

Not only is the regenerative paradigm the only mindset that will ensure your success while building your own permaculture prop-

---

The Problem with the Word "Sustainable"

I'll acknowledge that the most commonly cited definition for the word *sustainability* as well as *sustainable development* is from the 1987 Brundtland Report: "meeting the needs of the present without compromising the ability of future generations to meet their own needs."

Two important considerations influenced my use of the word *sustainable*, and subsequently *sustainable paradigm*, in this book. The first was that, everywhere I look, I find examples of the terms *sustainable* and *sustainability* being used by different people, organizations, and institutions in ways very different to what I think it means: to "meet the needs of the present without compromising the ability of future generations to meet their own needs." The concept of industrial-scale "sustainable beef," being one such example, or "sustainable pesticide use" being another. While many people are doing amazing work under the banner of sustainability, I often don't like to use the word because it has become too watered-down in our contemporary age.

The second issue is that the word *sustainable* comes from the word *sustain* plus the suffix -able. In dictionaries, the word by itself tends to be defined as the capability of a system to endure, maintain, or balance itself. In order to properly articulate the ideas here, I need to differentiate between the concept of something simply being *sustained* versus being actively stewarded and improved upon. I believe that *regenerate* is a better word in this regard.

erty, let me remind you that our planet is just one big complex system. So as your paradigm is fundamental to your own individual success, our collective paradigms are driving us rapidly towards environmental and societal collapse. Let's explore this.

Two Variations of the Downward Spiral

Suppose you are on your property, and a new pattern emerges; this could be a new plant, fungus, animal, or insect that is proliferating in the fields, or within your herds. We may be afraid that the organism is a threat to us (degenerative paradigm), or we are afraid that this "invasive species" is a threat to nature (sustainable paradigm). We therefore pull, plough, burn, fertilize, spray, trap, and/or poison that organism to destroy it. Or, perhaps we try to understand it; we isolate it in a lab away from its ecosystem and deduce all kinds of principles about what its function is. Not surprisingly, in its secluded state we discover it is even more of a threat to us than we thought, but at least now we better understand its weaknesses.

The practices used to destroy (or alter) that organism as an element within a complex system have dynamic and cascading effects. The result? Chaos and fragility ensue. Soil compaction, soil loss, biodiversity loss, population imbalance, drought, flood, and a myriad other patterns start to emerge that we could have never predicted. Suffering for us and others results in the form of extra work required to implement reactive practices. New pollution sources emerge, and more symptoms appear as the ecosystem attempts to self-organize and rebalance. But these new symptoms were not what we expected, so fear builds. Our flawed response is then magnified as we continue to make other poor decisions as systems managers. We ignore or eliminate feedback, we pay too much attention to details (instead of looking at patterns), we look for short-term "gains" and fail to think of the long-term consequences. We sacrifice resiliency for these gains, a good example of which is the practice of injecting cows with bovine growth hormone to increase milk production, despite a detrimental impact to their overall health. We also narrow the definition of "gain" such that it only includes those things important to capitalists and markets. Skip forward to Figure 5.3 (page 187) showing a permaculture farm on the left and an industrial farm on the right, and note that the only gain that the industrial farm is concerned about is measured in bushels per acre. In the end, we ignore environmental problems, resource constraints, and basic systems

> Western civilization has preferred love of death to love of life, to the very extent that its religious traditions have preferred redemption to creation, sin to ecstasy, individual introspection to cosmic awareness and appreciation.
>
> Matthew Fox,[15]
> author and theologian

> If the current predominant degenerative and sustainable paradigms and the resulting downward spirals can only lead to collapse, and if the root cause of these trajectories is our flawed reactions and response to complexity, then what is a better approach for living in a complex world?

science. We continue the pattern of *arrogance—extract—combat—suffer* (degenerative paradigm), or *guilt—conserve—control—survive* (sustainable paradigm). But more fragility and more unpredictability emerge, and the end result is that the downward spiral left in our wake erodes the well-being of all in its path.

Normally, the suffering incurred from living in disharmony with an ecosystem would cause an organism to either evolve or escape—else face extinction. Humans, however, found a loophole, or rather a way to delay the inevitable rock bottom. First, we domesticated animals to become our beasts of burden to pick up the extra slack and stop the chaos from consuming us. When the downward spiral continued, we started enslaving other humans to help manage our failing ecosystems so that they would continue to provide all our food, water, and energy needs. And finally, we discovered steam, fossil fuel, and nuclear power. These energy sources or **energy slaves**[16] continue to fuel an ever-expanding array of infrastructure all designed to free us of stewardship and make our suffering disappear. The terrible irony here is that our impeccable understanding of these inherently fragile, so-called complicated systems gave us the power to overwhelm the inherently antifragile complex systems we depend upon. And with the progressing degradation of our ecosystems, we have become ever more dependent on those same technological solutions. In short, it's a progress trap,[17] whereby the time the fragility and suffering is perceived it becomes bureaucratically, socially, or technologically impossible to do anything about it. With our heads down, we are kicking the can of suffering down a one-way street towards a cliff, all the while being completely oblivious to all the one-way and warning signs.

If the current predominant degenerative and sustainable paradigms and the resulting downward spirals can only lead to collapse, and if the root cause of these trajectories is our flawed reactions and response to complexity, then what is a better approach for living in a complex world?

Starting an Upward Spiral: Revere, Co-create, Design, Thrive

There is good evidence that we are currently in the midst of the sixth mass extinction event to have ever occurred on our planet and that humans are to blame.[19] While this is a rather depressing discovery, it also provides two of the most compelling arguments for hope and inspiration:

1. Our planet has already gone through five extinction events, yet still exists today.
2. Humans—not asteroids, comets, super volcanoes, coronal mass ejections, and pole reversals—are the cause of this event.

Let me elaborate on these points. To be considered a mass extinction event, at least half of all life-forms must be wiped out. In the largest of the previous five mass extinctions, upwards of ninety percent of all species were eliminated. Yet after each mass extinction, life and ecosystems have returned, each time more diverse and complex than before.

To take this insight further, at one point, planet Earth itself didn't even exist in this universe! And before the various particles of matter that would eventually coalesce into our planet could be ensnared by the gravitational pull of our sun, they had to come from the explosion of previously existing planets and suns that might have supported civilizations billions of years more advanced than our own. And the matter from those now dead celestial bodies emerged from the empty vacuum of space and time itself, as is now theorized by some physicists.[20] Regardless of the origin of our universe, either by divine conjuring or the mysterious evolutionary drive of life, every galaxy, sun, planet, organism, cell, atom, and electron exist today for the same reason. That reason is complexity. The same diverse, compounding, dynamic, and emerging characteristics of complex systems that were the starting point of the downward spiral we find ourselves in today are the very same qualities that are the basis, and hope, for the starting of an upward spiral.

If this seems like an impossible goal for humans, consider my second point that humans are the primary cause of the mass extinction happening at this moment. Or said another way, we are as powerful as the impacts, explosions, geologic events, and electromagnetic anomalies that killed more than ninety percent of all species on the planet in past extinction events! Think about the unbelievable intelligence, skill, creativity, collaboration, and perseverance it has taken humans to not only combat the most powerful force in the universe in an attempt to control it but also appear to actually be succeeding in doing so on a geologic time scale that is incredibly short.[21] On that note, I firmly believe that the reason we are eating, drinking, sheltering, and powering ourselves and our planet to death is not because humans are inherently stupid, lazy, and immoral; it is

> Most educated people are aware that we are the outcome of nearly 4 billion years of Darwinian selection, but many tend to think that humans are somehow the culmination. Our sun, however, is less than halfway through its lifespan. It will not be humans who watch the sun's demise, 6 billion years from now. Any creatures that then exist will be as different from us as we are from bacteria or amoebae.
>
> Sir Martin Rees,[18] professor of astrophysics

because of our inappropriate paradigms that have perpetuated into a downward spiral of shifting baselines with regards to assumptively "appropriate" and "normal" behavior. In some ways, we have been like a child who burns her hands on a light bulb because she was attracted to its glow. That glow was the promise of infinite growth without consequence that our scientific, corporate, political, and religious institutions have advertised to humanity since recorded history.

And what if burning our hands on the glow of progress was essential to the deepening and evolving of our philosophy of the patterns of the universe so that we would make meaningful changes to our practices? What if, as Carl Jung says, "No tree, it is said, can grow to heaven unless its roots reach down to hell"?[22] Regardless, if we arrived at our current predicament through innocent default or the malicious design of a minority of psychopathic, predatory ruling elites,[23] the solution to our problems and the end point if we fail to evolve remains the same: we must partner with complexity or face the inevitable rock bottom of the downward spiral that can only end in our own destruction.

Having already gone through five mass extinction events, it is safe to say that the planet will be just fine, but rather, it's the fate of humans that is in limbo. Luckily though, our fate is in good hands, because it is up to us. If humans, in our current Anthropocene, have as much destructive power as the events that brought about the extinctions at the end of the Ordovician, Devonian, Permian, Triassic, and Cretaceous periods, where more than half of all species were destroyed, think of the flourishing that could be created if we stopped fighting the force of complexity and instead used all of our power to embrace it. Humans can be just as regenerative as we are currently degenerative. But to do so we must evolve our collective paradigms, principles, practices and ultimately develop clear processes that facilitate the successful co-creation with the complex systems of this world. There is no telling what webs of life are possible if we focus on learning how to become a better species within them, rather than attempting to survey, suppress, and/or control them from above.

So, what is a step-by-step process for partnering with something complex and by definition is constantly and unpredictably evolving? Let's finally dive in!

> **The solution to our problems and the end point if we fail to evolve remains the same: we must partner with complexity or face the inevitable rock bottom of the downward spiral that can only end in our own destruction.**

Step 0: Inspect Your Paradigm   37

DEGENERATIVE PARADIGM

SUSTAINABLE PARADIGM

REGENERATIVE PARADIGM

FIGURE 0.2. The degenerative paradigm is based in the belief that humans are better than nature, and the sustainable paradigm is based on the belief that humans are worse than nature. In the regenerative paradigm, humans have the motto "We are nature," and they see themselves simply as members of a community of sentient life forms that are all working towards the collective and interdependent goal of well-being.

TAKOTA'S STORY | The Coen Permaculture Farm Upward Spiral

There is a dark, yet profound, joke among farmers that industrial agriculture has been nothing more than a fight to answer two questions: *How do I kill this thing that wants to live?*...and, *How do I keep this thing alive that wants to die?* The idea here is that there is a constant battle to kill weeds and pests while keeping unhealthy and weak crops or livestock alive. But what if farmers and land managers started to ask a different question? A question like this:

> *How do I create the conditions for all life to thrive?*

Answering that question has been the mission of my family and me at Coen Farm since 1988 when we first started our transition away from industrial to regenerative agriculture.

We have adopted practices like rotational grazing, cover cropping, agroforestry, and optimized tillage to improve both soil quality and quantity. We have restored historical wetlands and riparian areas and even built new ones to repair our water cycle. Every year since 2016, we have captured or infiltrated 45 million liters (10 million gallons) of snowmelt water via water harvesting earthworks such as swales, dams,

FIGURE 0.3. Coen Farm is an award-winning 100-hectare (250-acre) permaculture farm located in central Alberta, Canada. Photo taken in spring 2018.

Step 0: Inspect Your Paradigm   39

FIGURE 0.4. Coen Farm produces milk-fermented grain-fed pork, grass-fed beef, eggs, berries, and tea for their customers.

Credit: Takota Coen

dugouts, and wetlands. With only 1.1 million liters (250,000 gallons) of groundwater used for irrigation, livestock, and domestic uses, that means that we are capturing and infiltrating more than forty times the amount of water that we pump from the aquifer each year.

We have planted more than 30,000 fruit, nut, berry, and timber trees and installed hundreds of houses for birds, bees, and bats to increase natural biodiversity. Further, ongoing efforts have increased the diversity of our crops from just four or five species to well over one hundred, and we now favor perennial plants over annual plants to increase photosynthetic capacity over the course of the year. As a result of all this, species of birds, mammals, insects, amphibians, and reptiles not seen in generations are returning in droves. With the diversity of plants and proper land management, we now capture the sun's energy 365 days of the year on every square foot of land, unless it is covered by snow. When that snow melts, it is harvested by more than 3 kilometers (1.9 miles) of swales and a dozen dams, dugouts, ponds, and wetlands, where it can slowly spread and infiltrate back into an aquifer that is now showing signs of improvement after more than 100 years of decline. Springs have already reemerged after only 5 years.

What is most exciting is that, as we began to heal our land from over a century of abuse, we began to see improvements including increases in the nutrient density of our pork, beef, eggs, and other farm products in just a few decades. Every year our perennial customers attest to an increase in flavor and personal health, and nutritional tests performed by a researcher at the University of Toronto showed that Coen Farm pork, beef, and eggs had five to ten times the good omega fatty acids of equivalent foods on the market.[24]

We've achieved this without using any fertilizer, pesticide, antibiotics, vaccines, or any other synthetic chemical or drug since 1988, and only approximately 10 liters (~2.5 gallons) of fuel per acre, which we are working towards decreasing every year. On top of the food we raise for direct market to consumers, we also produce the majority of our own calories in our kitchen garden in an ever-increasing diversity.

Practices for Step 0. Inspect Your Paradigm

Although I expect that many of the readers who have picked up this book are likely already firmly steeped in the regenerative paradigm, it's not a bad idea to stop here and reflect before continuing. Remember that your permaculture property is a complex system, so in the end it really is about your ability to successfully partner with something that is evolving, emerging, and self-organizing. Any baggage from the degenerative and sustainable paradigms will hold you back from being successful in this endeavour. There are many effective ways to inspect and shift your own paradigms, so I offer the following two practices as simple suggestions that may help you spot a few gorillas hiding in plain sight.

Practice 0.1: Reflect on Major Influences, Biases, and Assumptions You May Hold

The first step is to critically interrogate your assumptions and the biases they perpetuate. Train yourself to admit defeat, that it's okay to be wrong, and just as okay (if not admirable) to admit it. Reflect on your belief systems, whether they be held within religion, cultural worldviews, political ideologies, or any other context entirely. Reflect on what your current beliefs are about questions like: *Where do we come from? Why are we here? Where are we going?* Review the scriptures, mythologies, manifestos, textbooks, novels, and/or any sources that shape your current beliefs about these and other philosophical questions while considering Table 0.1. Look out for statements, principles, and practices based on the belief that we are separate, better than, or worse than nature. You may want to cross these out. Next, look out for statements, principles, and practices based in the idea that "We are nature" and promote compassion and collaboration with all living beings. You may want to highlight these, or capture any particular statements on a separate sheet of paper. Hang on to this sheet of paper, it will be helpful to you later in Step 1, and consider sharing any insights with your accountability partner.

Practice 0.2: Study, Observe, and Contemplate Ecology

Spend time outside, get your hands in the dirt, plant a seed, watch it grow. Go for a hike in the forest, climb a mountain, hug a tree, drink from a mountain stream, watch a bird build a nest. Sit out late one night and watch the stars. Research local Indigenous teachings and seek out Elders. Pick up books from eco-philosophers such as Henry David Thoreau, Aldo Leopold, Wendell Berry, and more (and if you are interested in exploring some of the authors that have influenced our own thinking, seek out the authors that we quote throughout this book). Spend time in deep contemplation and meditation and immerse yourself in any outdoor settings that inspire you.

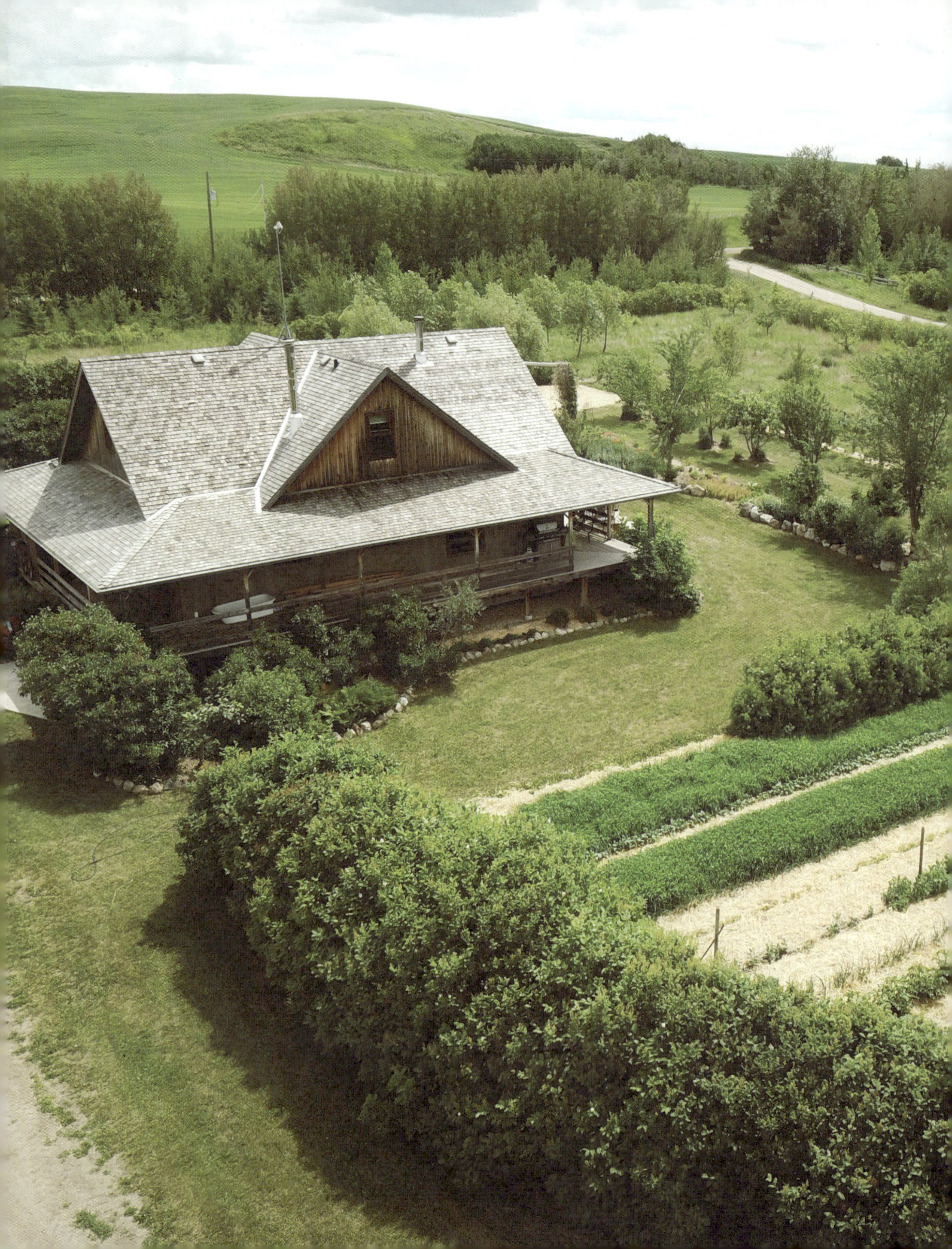

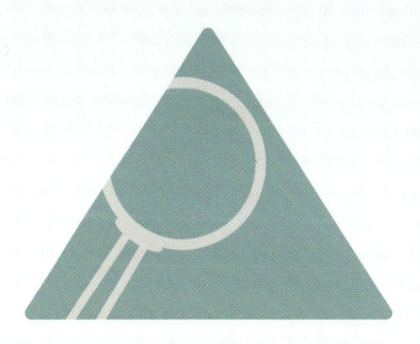

Step I: Clarify Your Vision, Values, and Resources

Clarify

clar·i·fy

\ ˈkler-ə-ˌfī

verb

To make clear by removing impurities.

Should I live close to the city or far away where land is cheaper? Do I buy bare land or a property with a house? One acre or one hundred acres? Hilly or flat? Should I build a passive solar greenhouse or moveable hoop house? Should I establish a food forest or an orchard? Chickens or pigs? When it comes to X and Y, which is the right one? And of course, my number one most asked question: Should I swale or **keyline**?[2]

All of these questions are a clear sign that the inquirer is stuck in the first struggle of permaculture: I don't know what I should do. And whenever I receive questions like this, I can't help a mischievous smile from taking over my face because I know I will soon get to reply: It depends. During the subsequent pause and nervous giggle from the questioner, I will often find a surface to write on to explain further. Across the top of the blank page, I write Infinite Possibilities, and in the center, I write the question *What to do?* with a big question mark. Then I draw three overlapping circles around it. Inside the first circle, I write, *What do you have?*; in the second circle, *What do you want?*; and in the third, *What is right?* When I have finished the Venn diagram, as shown in Figure 1.1, I explain that these are the three things that all these *What to do* questions depend upon, and that the solution is to clarify your vision, values, and resources.

It depends.

FIGURE 1.1. The infinite number of possible options that you could consider when building your permaculture property can be significantly filtered down if you take time to build a Venn diagram that constrains *What you want*, *What you have*, and *What is right?*

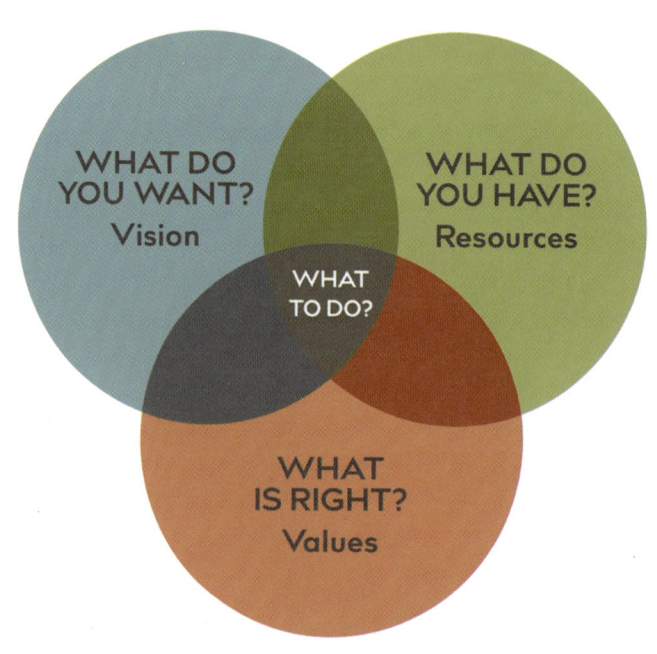

Credit: Shauna Fidler

The good news is that permaculture already has most of the tools and ideas to help us in this process of elimination. These tools and ideas just need to be organized and applied in a different way than they have before. In particular, the quintessential **needs and yields analysis** exercise of permaculture design—which involves selecting an element in the design and listing what it requires and what it produces—is rarely completed for the central element of our systems: ourselves. As you will see, when we apply this method of design to ourselves, it helps us to solve for the first two sets in the Venn diagram: *What do you want?* and *What do you have?* Further, guidance with the final set, *What is right*, can be found in none other than the **three permaculture ethics**: earth care, people care, and future care.

This chapter expands upon these existing concepts within permaculture to help you solve your own Venn diagram and overcome the struggle of *I don't know what I should do*. In what follows, you will learn:

- A framework that will provide the foundation for a complete and holistic inventory of your resources.
- How the regenerative, sustainable, and degenerative paradigms each influence the potential answer to the questions: *What is right?* and *What do I want?*
- How the three permaculture ethics, earth care, people care, and future care, form an excellent foundation for morality.
- A simple thought experiment that you can use to determine if you are doing the right thing.
- Why you and your partner(s) are not on the same page and how to fix that.
- How to effectively de-escalate situations that inevitably arise when you and your partner(s) discover incongruencies in your vision and values.
- The four biggest mistakes people make when clarifying what they want.

What Do You Have?

Before deciding *What to do* for even small projects, let alone designing and developing land, it is clear that having a clear inventory of what resources are available to you to pull off the job would be helpful. The two things that come to mind first are usually your financial and material resources. But there's actually much, much more to

> If one does not know to which port one is sailing, no wind is favorable.
>
> Lucius Annaeus Seneca,[1]
> Roman Stoic philosopher

TABLE 1.1. Your Personal Resources. Each of the eight personal resource categories, common currencies, a few examples, a description of how that category may increase well-being, and some existing inventories that you may keep. Adapted from Ethan Roland and Gregory Landua's *Eight Forms of Capital*, regenterprise.com

Personal Resources	Common Currencies	Examples	Usefulness in Increasing Well-being	Common Existing Inventories
Financial	Money	Equity or debt in the form of income, savings, investments, grants, loans.	Our primary tool for exchanging resources between one form and another. Apart from its ability to make other resource exchanges efficient, it has no intrinsic value.	Accounting records, bank account statements, investment portfolios, loan and credit card statements.
Material	Materials, nonliving "natural" resources: stone, metal, fossil fuels, etc.	Buildings, infrastructure, equipment, tools, computers, technologies, etc.	The "stuff" we own or use. When it comes to well-being, material resources are usually a means to an end.	Depreciation schedule, asset list, insurance addendum(s), photos, equipment lists, maintenance records, receipts.
Living	Carbon, nitrogen, water.	Animals, plants, water, and soil.	The animals, plants, water, and soil of our land—the true basis for life on our planet.	Property/land title, crop log books, livestock log books, gardening journals, plant inventories, soil carbon tests, water tests, biodiversity studies, etc.
Social	Connections	Equity or debt in the form of influence, relationships, favors.	A person or entity who has "good social capital," can ask favors, influence decisions, and communicate efficiently. Humans are social beings and often attribute significant well-being (or suffering) to their social resources.	Address book, customer/supplier lists, social media platform contacts, email contacts.
Spiritual	Prayer, intention, faith, karma.	Any way that you might measure your mental state (balanced, peaceful, etc.) or any measurement of spiritual attainment.	The *intent* of what you do and what drives you or guides your actions. This is why you get up every morning and is intrinsically tied to well-being.	Diary, journal, photo albums, personal art portfolio, etc.

TABLE 1.1. (cont'd)

Personal Resources	Common Currencies	Examples	Usefulness in Increasing Well-being	Common Existing Inventories
Experiential	Action	Things you have done or have experienced doing. Embodied knowledge.	Your embodied knowledge (i.e., skills) that can both support, and be tied intrinsically to your definition of well-being.	Resume, achievement certificates, awards.
Intellectual	Ideas, knowledge.	The "knowledge" assets like words, images, and intellectual property (patents and copyrights).	Intellectual assets and ideas can both support, and be tied intrinsically to your definition of well-being.	Computer hard drive, content platforms for videos, blogs, podcasts, portfolios, list of books in your library.
Cultural	Song, story, ritual.	Arts, celebrations, traditions, taboos, languages, myths.	When individuals form a family, village, city, bioregion, or nation and begin to exchange various forms of resources, patterns begin to emerge. Cultural resources are an expression of our internal and external processes as a community.	Photo albums, family videos, event calendars.

one's resource base than the money in your bank account or the tools and supplies in your garage. Before you can get the best answer to the question *What to do*, you need to have a holistic inventory of all the resources that you own or have access to, that will help, or perhaps even hinder, you.

The Eight Forms of Capital,[3] developed by permaculturalists Ethan Roland and Gregory Landua, is just such a holistic framework for expanding our definition of resources. The framework includes categories such as living, social, experiential, intellectual, spiritual, and cultural resources, in addition to the more common financial and material resources. Table 1.1 describes them and, for each one, shows common currencies, its usefulness in the ultimate goal of increasing well-being, and common existing inventories.

A **personal resource inventory** involves listing *all the things* within these eight categories that could be of use or leveraged to accomplish one's goals, and also any negative resources or debts within those resources. For example, you could owe the bank money, a debt to your financial resources, or a friend lots of favors, a debt to your social resources. I remember having three major insights when I first took the time to complete this exercise. The first insight was how "rich" I truly was. Before, if someone only used my income tax returns as a metric of my wealth, it would appear that I lived just above the poverty line! Now, with a framework to holistically communicate the other valuable resources I owned or had access to, that same person would see me for the millionaire I had always felt like. Best of all, this wealth was all tax-free, and immune from the volatilities of a fiat money system! The second epiphany was that I now had a clearer sense of the destination that I *actually wanted* in addition to the newfound resources that were a means to get there. Dying with a stuffed garage and bank account had never appealed to me. But investing in my living, social, spiritual, experiential, intellectual, and cultural resources became strong motivators. This was true well-being. My third insight was that my struggle to figure

FIGURE 1.2. If you don't know what you have, no matter how much, it will never be enough.

Credit: Jarett Sitter

out what to do started to seem less complex. In the same way that water infallibly flows towards the lowest point in the landscape, my decisions seemed to be drawn intuitively towards a new focal point that was an expanded definition of well-being.

Thinking back to Figure 1.1, building a personal resource inventory helped me to draw the *What do you have?* and *What do you want?* circles to constrain all the infinite options in my life. In doing so, my struggle to figure out *What to do* began to drain away.

I've since included a resource inventory exercise in every course that I teach on permaculture design and start every consultancy with asking my clients to complete a resource inventory of their own. This exercise takes several hours, but it is a worthwhile and empowering effort as it provides you with a sense of wealth and well-being you had no idea you already had. A little later, in Step 2: Diagnose, you will be further investigating this resource inventory for strengths, weaknesses, opportunities, and threats, and in Step 3: Design, you'll discover that this inventory will also help you to avoid missing opportunities, as illustrated in Figure 1.2. There, you'll see three designers trying to climb over a wall, and two of them are simply standing on a pile of ladders not realizing that the pile itself contains the solution to get over!

When you create your personal resource inventory, you may want to choose a format that is easy to navigate and keep up to date. A digital spreadsheet or three-ring binder with tabs for the eight personal resources both work well. And rather than build it from a blank sheet of paper, you can simply look at the column "common existing inventories" in Table 1.1 and file a copy of the existing list(s) or statements under the appropriate tab. For instance, store your most up-to-date financial statements or bank summaries under the tab "Financial Resources," a list of the important people and social connections on a page behind a tab called "Social Resources," and so on. But you'll likely find that the existing lists are incomplete, and you'll also need to do some brainstorming so that you can capture as complete a picture as practical relative to your project.

Practice 1.2 (at the end of this chapter) is where I'll step you through how to create your personal resource inventory in slightly more detail. But now let's take a deeper look into the other two problem sets, *What is right?* and *What do you want?*, as we continue to solve our Venn diagram for *What to do*.

FIGURE 1.3. When creating your resource inventory, a digital spreadsheet or three-ring binder works well, depending on your preference. Both can have tabs for the eight personal resources to help you get organized.

Philosopher-gardeners, or farmer-poets, are distinguished by their sense of wonder and real feeling for the environment. When religions cease to obliterate trees in order to build temples or human artefacts, and instead generalise love and respect to all living systems as a witness to the potential of creation, they too will join the many of us now deeply appreciating the complexity and self-sustaining properties of natural systems, from whole universes to simple molecules. Gardener, scientist, philosopher, poet, and adherent of religions all can conspire in admiration of, and reverence for, this earth.

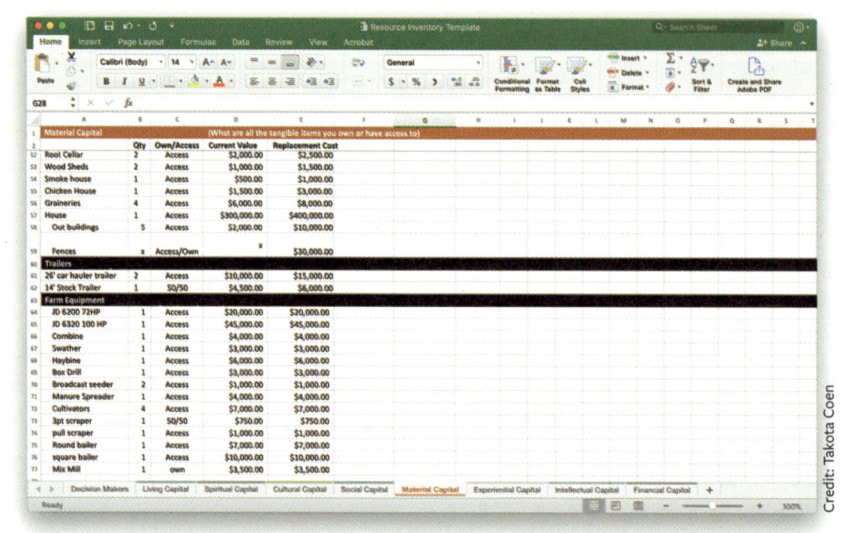

Bill Mollison[4]

What Is Right?

So what is the right thing to do anyways? To put this in terms of the three paradigms, the degenerative paradigm's answer to what is right is often predicated on the belief that humans have the divine or evolutionary dominion to plunder the Earth for their own benefit. They tend to have a selfish perspective—*How do I get more goods?*—and are trying to figure out how to get those goods as fast as possible. Of course, their definition of "goods" is limited to money and financial resources. Humans in the sustainable paradigm attempt to be selfless: *How do I do less bad?* However, many members of the sustainable paradigm still engage in the same looting of the planet, then acknowledge their wrong actions with justifications or diversions of a percentage of their spoils to their church or some greenwashing environmental charity. While these two viewpoints ferociously debate the answer to *What is right?* amongst themselves, those in the regenerative paradigm prefer to consider the question differently altogether by asking: *Why should you care if you do the right thing?*

There are at least two reasons you should care about doing the right thing. The first is because it minimizes your own suffering. The second is because it maximizes your own well-being. If you look back to Figure 0.1, you will remember that the downward spiralling effect of suffering emerges when the complexity of living systems is not respected. Conversely, there are unimaginable, unlimited, and possibly even infinite increases in well-being for every element in the system when it is. Or as Ojibwe writer Richard Wagamese put

FIGURE 1.4. While those in the degenerative and sustainable paradigms waste their time arguing over the shape of the shadow that they see, those in the regenerative paradigm recognize that we must take responsibility—and action—or face the consequences.

it, "We live because everything else does."[5] That is why our values need to reflect our entanglement with the well-being of all organisms, and perhaps even geological features. Many will argue that this concept of morality and its link to the well-being of all is illogical: How can you value the lives of plants, animals, and even mountains as equal to that of a human? Anthropologist Wade Davis provided the following rebuttal:

> A young kid from the Andes who's raised to believe that a mountain is an Apu spirit that will direct his or her destiny will be a profoundly different human being, and have a profoundly different relationship to that resource or that place than a young kid from Montana raised to believe that a mountain is a pile of rock ready to be mined. Whether it is an abode of a spirit or a pile of ore is irrelevant. What's interesting is the metaphor that defines the relationship between the individual and the natural world.[6]

I am brought to tears of joy imagining future generations enjoying the fruits of the trees I have planted at great financial expense that I know will offer me no return on investment in my lifetime. I do this not out of fear or hope of cosmic retribution or reward, nor in conspiracy with my DNA to perpetuate myself. I do it because of the increased sense of experiential, social, and spiritual resources that improve my well-being in the present, second to none. Just as the Greek proverb asserts, our society will grow great when old men plant trees whose shade they know they shall never sit in. This is the right thing to do not because of a sense of duty, or fear, or sacrifice, or guilt but out of enlightened self-interest that comes with the expanded definition of well-being as being more than just money and things. This means that, when viewed through the lens of the regenerative paradigm, the answers to the questions about *What is right?* and *What do I want?* converge because of the inseparable link between ethics and personal suffering and well-being in this life. And as per the prime directive of permaculture: "The only ethical decision is to take responsibility for our own existence and that of our children."[7] The minimum definition of *What is right?* is to do only those things that are good for oneself and all our relations and that allow all others the same opportunity or better into the future. Our ethics must be earth care, people care, future care.

Just as the Greek proverb asserts, our society will grow great when old men plant trees whose shade they know they shall never sit in.

A Moral Landscape and a Veil of Ignorance

Over 2,300 years ago, the Greek philosopher Aristotle attempted to answer the question *What is the meaning and purpose of life?* The answer he proposed[8] is **eudaimonia**, a Greek word that roughly means human flourishing and is often translated into English as "happiness." The trouble with that translation is it implies that eudaimonia is an emotion or feeling. However, Aristotle himself described eudaimonia as "both living well and doing well"; taken literally, the person who embodies the ideals of eudaimonia, is a eudaimōn, or good soul. With that, I much prefer translating eudaimonia to well-being.

I'm sure that you have many reasons for wanting to build your own permaculture property, or help others build theirs. Regardless, throughout this book, I'm going to explain why I think the Greek concept of eudaimonia, or well-being, as I'll refer to it going forward, is, or at least should be, the fundamental answer to the question: Why build a permaculture property? And I don't just mean well-being for ourselves, our neighbors, our communities, and our fellow humans. I mean well-being for all life on this planet now and into the future. Only when we collectively focus on the goal of well-being for all will we humans be able to truly thrive.

This summum bonum (the highest good) should be our goal; in other words, to become a eudaimōn and live virtuously. Sam Harris offers contemporary insights on exactly how to do this: "The moment one begins thinking about morality in terms of well-being, it becomes remarkably easy to discern a moral hierarchy across human societies."[9] The moral hierarchy that Harris refers to is a moral landscape, whereby just like the topography of the terrain has peaks and valleys and a gradation between them, so do our concepts of suffering and well-being. If you are standing on one point in a landscape and looking across to other peaks, you can claim that some are higher and lower than others, but because you cannot see the entirety of the terrain, you cannot claim which is the highest and which are the lowest.

In the same way, it is possible to compare two paradigms, principles, practices, or processes that are side by side and discern which is *better*, but, because of complexity and the limitations of our paradigms, it is impossible to say which one is *best*. The real importance of this definition of well-being is not in the fixed definition but

> **Only when we collectively focus on the goal of well-being for all will we humans be able to truly thrive.**

TAKOTA'S STORY: Two Paths to the Same Cliff

In my early twenties, I was a classic capitalistic consumer stuck in the degenerative paradigm. I monitored my growing bank account balance obsessively. I owned a jacked-up pickup truck, a dirt bike for weekend trips to the trails in the mountains, a street bike capable of highway speeds of 300km/hr (180 miles/hr), and all three of them were equipped with the loudest mufflers money could buy to draw maximum attention to myself. After being unable to fill the void of meaning in my life with money and things and after some shameful reflection about how I was behaving in light of the obvious suffering in the world, I sold all my toys, switched to the sustainable paradigm, and turned into a stereotypical "tree-hugging hippie." I started marching in protests and regularly berated family and friends for not doing more about doing less bad. Recycling miniscule pieces of plastic and depriving myself of simple pleasures like hot showers became my new obsession. I truly believed that to want or have anything was not right.

Having given up eating meat because of my desire to not be complicit in the suffering of conscious creatures, I was dismayed to learn during a one-year apprenticeship with a herbalist that the evidence for the sentience of plants was also undeniable. My own research[11] and reflection, combined with personal spiritual experiences, made it clear to me that plants also had the ability to feel, communicate, make decisions, remember information, and show preferential cooperation with their species, even their kin! They were trying to survive and thrive, and arguably, increase their well-being, just like animals. Just like me! This presented a dilemma: If I couldn't eat animals, and I couldn't eat plants without causing suffering, what the hell was I going to eat?

This extreme guilt for even being alive grew each day until eventually I reached rock bottom and my internal conflict erupted during a family get together. I had arrived early to my parents' house to find my mother in the kitchen, happily preparing one of her delicious family meals. Then I noticed, in the center of a beautifully set table, a small vase of lilac flowers she had just picked from the yard. I lost control of my unconscious self-hatred, self-righteousness, and self-pity and projected it onto her in a murderous rage, screaming, "It was unnecessary to take the life of an innocent plant that we were not even going to eat!" Later that night, after an uncomfortable family meal, I had a dream that turned my sustainable paradigm on its head.

In my dream, I was riding upon the back of a radiant white cow towards the center of an old village. People lined the streets dancing, singing, and happily casting garlands of flowers upon me and the cow's magnificent horns. Upon reaching the gathering point of what appeared to be a ceremonial altar, a man helped me to dismount. Standing, facing the cow, I admired its calm demeanor before I began surveying the large crowd of smiling people that had emptied the streets and now stood waiting in gleeful anticipation in front of the altar. Absolute joy filled every cell in my body. But this sensation of bliss was replaced with rage when I finished turning full circle to find the white cow dead on the ground in a red pool of its own blood that was gushing from its severed neck. I spun around furiously, finger cocked, to denounce the people for their immorality, just as I had scolded my mother earlier that day. But to my surprise, I found the entire crowd kneeling upon the ground with heads bowed in deep reverence and gratitude for the life that had just been taken. I awoke from the dream in a cold sweat with the answer to my question of what the hell am I going to eat etched into my memory: It is not what you eat, it is how you eat it.

With every breath, step, or bite, millions of organisms are annihilated by my immune system, crushed under foot or dissolved in our stomachs. Everything, even the planets and our universe, is born, grows, then fades away so that something new can repeat the same pattern. Death literally feeds the creation and the complexity of life. To fear or attempt to control death is another core similarity in between the degenerative and sustainable paradigms.

Everything gardens.[12] Every living being interacts with its environment to facilitate its own needs. It is how we garden that matters. There is and always will be suffering associated with life and death. But it is possible to do things that minimize suffering, maximize well-being, and use both as an indicator for what is right. Or as Wendell Berry put it: "We cannot live harmlessly at our own expense; we depend on other creatures and survive by their deaths. To live, we must daily break the body and shed the blood of Creation. The point is, when we do this knowingly, lovingly, skillfully, reverently, it is a sacrament; when we do it ignorantly, greedily, clumsily, destructively, it is a desecration."[13]

I consciously work to provide the chickens, pigs, cattle, and every other animal, even the plants, on my farm with such an excellent life and a compassionate death that I myself would agree to have our roles reversed. In this context, it didn't surprise me to learn that there is even evidence that some animals domesticated themselves.[14] Regardless, I know both that my animals choose to remain on my farm today and that I, myself, would prefer to live as an animal on this farm where I have all freedom as in the wild but with none of its perils.

To ensure that these creatures actually want to be on my farm is no easy task. It requires constant attention to their needs and their yields. To their well-being. It requires "husbandry." This is what permaculture really is: it is a marriage between oneself and all beings. And when this marriage is lived knowingly, lovingly, skillfully, and reverently, it is, in and of itself, the greatest source of well-being I know. This is my new paradigm. And it has been as regenerative for me as the ecosystem, the marriage, I belong to.

rather the iterative and ongoing process of asking, and then defining *What is right*.

When it comes to evaluating how to increase well-being for all, there is a fascinating thought experiment coined the veil of ignorance.[10] In this exercise, a hypothetical future society (i.e., system) is designed, in the present moment, by its members. They would make rules and policies about things such as taxes, education, homelessness, and access to resources. However, none of the designers know which role they would get upon entering that new society. For instance, you could enter as a homeless drug addict, or as a starving child in Africa, or as a wealthy business woman. I think it's evident that the world would be a more just and better place for all if societies were designed this way. When it comes to being designers for our own complex systems, in order to truly propel the upwards spiral, this thought experiment must be expanded to include all nonhuman organisms as well. A truly ethical system is one that you would wish to be any element within, regardless of your function within it. In addition, we must recognize that our work as designers is never done as we constantly engage in an iterative and ongoing process of defining what is right. In fact, it is the defining more than the definition of *What is right?* that is truly important.

> A truly ethical system is one that you would wish to be any element within, regardless of your function within it. In addition, we must recognize that our work as designers is never done as we constantly engage in an iterative and ongoing process of defining what is right.

What Do You Want?

To make the benefits of clarifying your vision and values less abstract and more personal, I like to play an alcohol-free version of the party game Never Have I Ever at the start of my workshops on goal-setting and decision-making. I start by having the entire class stand up and explain that they are to sit down only if they can truthfully say that they have never experienced any one of the following:

Never have I ever…
- Made a decision about what to do that I regretted terribly.
- Lost sleep agonizing over a decision about what to do.
- Felt lost and directionless about what to do in life.

Never have I ever…
- Had a fight with a partner over a decision about what to do.
- Felt dread about needing to make a decision about what to do with a partner.
- Ended a relationship over a disagreement about what to do.

Never have I ever…
- ▸ Felt that our global leaders make reliably terrible decisions about what to do.
- ▸ Witnessed the terrible global consequences of even talking about what to do.
- ▸ Felt like humanity is and always has been divided when deciding about what to do.

I have yet to play this game where a single person sits down. Take a moment to let the individual, community, and global implications about our struggle to figure about what to do sink in.

Fortunately, the solution to the problem of not being on the same page about what to do is relatively simple. Create one!

Get on the Same Page

In addition to the resource inventory that you created to clarify what you have (Figure 1.3), create a single physical page that clarifies what you want and what is right. I call this your vision and values one-pager. Figure 1.5 provides a template for a vision and values one-pager that uses just nine sentences. The first sentence states your fundamental values (which you can choose to base on the permaculture ethics of earth care, people care, and future care), and eight others are your vision statements. I recommend that you write one for each of the eight resource categories: living, social, experiential, intellectual, spiritual, cultural, financial, and material. Each vision statement defines the optimal expression of that resource with respect to your greater well-being. Once completed, this document will not only get you and your fellow decision-makers on the same page, it will also act as a personal coach to motivate you when you are on the right track or as an arbitrator that can help guide your actions if you get lost.

A doubtful reader may be thinking, "Why should I clarify what I want? I know what I want!" I've also had a fair number of clients initially balk at my insistence on completing this step. But the reality is that, often, what we think we want is clouded by our paradigms and limited by a narrow vision of our personal resources. Taking the time to complete a resource inventory and build a vision and values one-pager gets your ideas outside of your own head and into the light of day where they can be inspected and shaped with the

> A people without an agreed-upon common basis to their actions is neither a community nor a nation. A people with a common ethic is a nation wherever they live. Thus the place of habitation is secondary to a shared belief in the establishment of an harmonious world community.
>
> Bill Mollison[15]

Values

To live in a way that is good for myself and all my relations and allows all others the same opportunities or better into the future.

Vision

Financial. I have a diverse surplus of financial resources and ongoing reliable income generation that is aligned with my vision and values.

Social. I have beautiful, supportive, loving, honest, and meaningful relationships within a thriving and stable community of self-motivated, mature beings that value their self-worth and are mindful of the feelings, rights, and human dignity of others.

Material. I own or have access to the minimal amount of ethical possessions that support me in being resilient, secure, and safe and/or bring me joy.

Living. I am in excellent health, stewarding a serene environment that is ever increasing in abundance.

Intellectual. I contribute my ideas towards the collaborative, ever-growing pool of accessible and meaningful information that is making the world a better place.

Experiential. I continue to share and learn a mastery of the skills that support my vision and values.

Spiritual. I am confident, peaceful, and in a balanced mental state where I have the freedom to self-direct my time, pursue my passions to their fullest potential while practicing service to others, gratitude, and forgiveness.

Cultural. My family and community rituals, traditions, art forms, and social patterns inspire all to create beauty and celebrate the shared values of earth care, people care, and future care.

FIGURE 1.5. This is an example of what your vision and values one-pager could look like. The values statement at the top can be based on the permaculture ethics (earth care, people care, future care). Below are eight vision statements, one for each of your personal resource categories.

clarity of distance. This will help you to figure out what to do in a way that leads to greater well-being for all. And it will lead to better and easier decision-making, with more alignment, less conflict, and increased peace of mind. It is an absolute prerequisite to successfully building your permaculture property.

Ranking Your Vision Statements

After their vision and values one-pager is complete, one of my favorite exercises to do with clients is to have them consider how close their current reality is to the vision statement they have just written. One by one, I encourage them to review the individual statements for each living, social, experiential, intellectual, spiritual, cultural, financial, and material resource and then ask themselves, *How close am I to achieving this vision?* Often people feel that they are doing really well in one or two of the personal resource categories but are falling short, sometimes substantially, in other areas. It can also be really interesting when partners see the gap (or lack of a gap) differently—which then spawns subsequent conversation and insight. Again, it is the reflection not the writing that has real value here.

The last step of this exercise is to agree upon and then actually rearrange the order that the vision statements appear on your one-pager. Those statements where the gap is largest go at or near the top of the page. Note that this intentional ordering is important and influences other exercises in the remainder of the 5SP Process—more on this later.

Imagine If…

After seeing dozens of completely different versions of the vision and values one-pager that my students, clients, and colleagues have built, I have yet to read one that I would not have wanted to put my own name on, with, at most, a few small tweaks. It really is amazing that, when we take the time to stop and reflect, we really do all want the same, and we all know what is right. We really are all on the same page—we just don't know it. It often leads me to wonder what might happen if everyone on our planet took the time to do these simple exercises. Jared Diamond summarized his study of collapsed societies: "Two types of choices seem to me to have been crucial in tipping the outcomes towards success or failure: long-term planning and willingness to reconsider core values."[16] With that in mind, our

> It really is amazing that, when we take the time to stop and reflect, we really do all want the same, and we all know what is right.

own downward spiral is surely a symptom of our societies' lack of clarity about what our resources, vision, and core values truly are.

So take a moment and imagine if our cumulative perception of the world was driven by the belief that resources to survive and thrive were under the right care, abundant and infinite, not that they were scarce, or merely enough. Imagine if every human on planet Earth saw themselves as a strand in the web of life working towards the common goal of well-being. Imagine how the power and creativity of the entire human race could be purposefully channeled into an unstoppable and infallible force for the advancement of all life. Imagine the world we could collectively create if we clarified our vision, values, and resources.

The remainder of this chapter focuses on some tips for working with your partner(s), resolving interpersonal conflicts, and developing shared vision and values statements, but first a brief recap on what we've covered so far. To best answer the question *What do you do?*, you need to first answer *What do you have?*, but with an expanded definition of resources. Next you need to clearly define *What do you want?* and put it down on paper. The last question, *What is right?*, is none other than the three permaculture ethics: earth care, people care, and future care. By clearly articulating and thinking through the answers to these questions, you'll find that the infinite number of *What to do* possibilities dissolve, or at least are significantly narrowed down, while the best unique path for you starts to come into focus.

Walking Through a Field of Landmines Blinded by a Scarf

If you or your project partner(s) are not used to talking about what is most important to you, if there is existing tension in your relationship, or if any of you are uncomfortable challenging your own biases, then you and your team are walking into a warzone of emotions that can lead to stressful arguments, wasted time, and decades-long grudges.

Although I don't claim to be an expert in the resolution of interpersonal conflict, I have learned enough for many of my clients and students to joke that I'd make a pretty good marriage counselor! Here I'll share the four most useful of those tricks via an analogy I call "walking through a field of landmines blinded by a scarf."

Step 1: Clarify Your Vision, Values, and Resources   61

Trick #1. Take off the scarf and see the landmines
A large number of people are completely unaware of their emotional needs, as well as the emotional needs of others. Using insight from brain scans, neuroscientist Dr. David Rock[17] developed the acronym SCARF to identify five social needs that are especially sensitive to being triggered and causing emotional explosions: status, certainty, autonomy, relatedness, and fairness. In Figure 1.6, these unmet SCARF needs are represented by the landmines strewn about as well as the figurative scarf that kept you blind and unaware of them. Contrary to **Maslow's hierarchy of needs**, status, lack of certainty, autonomy, relatedness, and fairness negatively impact our ability to survive and thrive as much, if not more than hunger, thirst, rotten food, and real physical pain or fear. In fact, these SCARF needs conjure up responses that can be detected in the same regions of

FIGURE 1.6. When working with other decision-makers, remember that there exists a no-man's-land between each of you that is filled with both of your emotional landmines. To navigate this minefield safely, you need to stop being blinded by a scarf!

Credit: Jarett Sitter

the brain as if there were a physiological stimulus. Just consider the following scenarios that you've likely experienced:
- Someone embarrassed you in public, challenging your status and causing your entire body to flush red or your blood to boil.
- Your future felt uncertain, and the gut-wrenching anxiety and lack of sleep this caused.
- Your ability to act as a free person was denied, like when someone took a toy away from you as a child.
- Someone you loved made you experience any of the previous emotions and how alone you felt as if no one could understand or relate to you.
- You were treated unfairly and the stomach churning and acrid taste of disgust you felt from this betrayal.

Thankfully, the opposite to these emotional triggers is also true. If you reflect what you like about your favorite friend and why you like them so much, it is likely because they improved your social resources and your well-being not by providing you with food, water, and energy but because they increased your sense of status, certainty, autonomy, relatedness, and fairness.

When working with your partner or other decision-makers in your project, remember that there exists a no-man's-land between each of you that is filled with both of your emotional landmines. To navigate this minefield safely, you need to stop being blinded by a scarf!

Trick #2. Be aware of the emotional explosions

If you happen to trigger someone's emotional landmines, which will inevitably happen, the second trick is to be aware of the two kinds of explosions: silent or violent. I'm sure you've experienced both before. The first is when someone clams up, and the second is when they lash out. Regardless of the nature of the explosion, if your partner has an endothermic or exothermic reaction, it's for the same reason: they feel threatened and are trying to escape or defend themselves. Both are a signal that your partner feels like something they value deeply is in danger and they are acting in a way to protect it.

And the larger the explosion the more unsafe they feel. Your partner may even respond with a level of violence or silence that sends you reeling from the impact, setting off your own emotional landmines. You may have stepped on a landmine by accident, or someone may have misunderstood you and put up their guard or thrown down their gloves in a way that triggers your own emotional explosion. But just imagine how hurt they must have felt to respond that way! From experience, I know that this kind of emotional intelligence is extremely hard to maintain. But the alternative is an argument that can turn into a tit-for-tat chain reaction of explosions that can only lead to the mutually assured destruction of your relationship. You don't have to condone someone's actions in order to have compassion for why they might have done something.

Trick #3. Disarm the bomb

The good news is that, once you start to notice these emotional explosions, you will realize that even the largest landmine has a few second fuse before damage is done by the other person putting up their guard or throwing down their gloves. If you are quick, and you know to cut the blue wire and not the red one, you can disarm the bomb with just one or two sentences that are designed to help the other person feel safe.

The formula for these sentences[18] is as follows:
1. Label the cause of the other person's silence or violence.
2. Acknowledge the fact you stepped on one of their emotional landmines.
3. Make clear your commitment to the other person's well-being.
4. Sometimes it helps to throw in an apology, even if it was just a misunderstanding.

Let's go through an example of this for each of the SCARF landmines:
- Status: I'm sorry, I didn't mean to belittle you, my goal was to see if there was any way I could help you with X.
- Certainty: I didn't mean to overwhelm you with the details, I know X is important, and I want to make sure we are on the same page.
- Autonomy: I'm sorry, I shouldn't have assumed you wanted to do X. How can I make this work best for you?
- Relatedness: It seems like you are distant, is there something you want to talk about?
- Fairness: I'm sorry, that wasn't fair of me. How can I fix this?

You can even use this formula if someone sets one of your emotional landmines: "I'm sorry, I didn't mean to attack you, I just felt threatened by what you said and was being a jerk."

Note that it is critical that these statements must be sincere. Status, certainty, autonomy, relatedness, and fairness are all needs developed over eons of evolution, and we are all remarkably good at detecting phonies. If you don't mean it, don't say it.

Trick #4. Fill in the craters

If, for some reason, you aren't able to disarm the landmine in time, or if you are like me and learned about the importance of these SCARF needs when emotional damage has already been done, then you will need to learn how to do some emotional restoration work. Any expression of violence or silence that goes unaddressed will leave a crater that can be stumbled into at some point in the future. Think about the time you got into a major argument with someone over something trivial like crumbs on the counter.

The good news is the reclamation work follows the same formula above, except that apologizing is a necessity now. This kind of reclamation work might take months or even years to fill in all the existing craters before any progress can be made in clarifying your vision and values. But in the same way you have to regenerate the health of your soil before you can get a productive garden, you have to regenerate the emotional health of yourself and your partners before anything good will grow.

> By not being blind to emotional needs, being aware of emotional explosions, learning how to disarm emotional landmines, and fill in emotional craters, you can turn even the most devastated war-zone land into a flourishing ecosystem.

Because, after all, what good is a permaculture property if no one can stand to share it with you?

Be Careful What You Wish For

Clarifying your vision and values may sound easy enough, but in my experience, there are some common mistakes that most people make. All of these are the result of the cognitive biases we all have. One of the best things about writing down your vision and values is that it allows you to take off the eyeglasses through which you have been viewing the world and inspect them for cracks. By bringing your unconscious assumptions into your conscious awareness, you can then reaffirm or reject them.

The four mistakes I commonly see are: (1) focusing on what over why, (2) using the forbidden word "will," (3) writing from the mindset of a degenerative or sustainable paradigm, and (4) becoming paralyzed by perfection.

Below I go into each of these mistakes in more detail and provide some tips on how to avoid these errors when creating your own vision and values one-pager. Also, in the upcoming story, Buckets of Well-being, Takota shares a tangible example where he used several tricks from "walking through a field of landmines" to help a client struggling with these common mistakes.

Mistake #1. What over why

When creating your vision and value statements by yourself or with a partner, leave out *the what* and *the how* and focus on *the why*. Try and filter down to what is fundamentally important to you, and that means staying away from specific quantitative things or actions. This is different from defining discrete and actionable goals (for instance: I want chickens), and instead we are looking for the reason why you think it is important to have chickens.

After you author your first draft, repeatedly ask yourself the following two questions to help you arrive at the heart of what it is you really want:

1. What is it about this that is so important to me?
2. Is there anything else that provides this same kind of well-being in my life?

Doing this for the goal "I want chickens" would likely lead to statements about food security, and even interdependency with the natural world. You can stop asking these questions when they no longer lead to deeper insights about what you really want. And if you are like me, you will also get goosebumps when you get there!

In the subsequent steps of the process (diagnose, design, implement, and manage), we'll start evaluating *the what* or *the how*. But not right now. At this point, remember to simply focus on why and not what.

It is often those with the strongest convictions about their own goals and destiny that are bound for the biggest struggles and disappointments in land and in life.

Ben Falk,[19] author and permaculture teacher

Mistake #2. Using the forbidden word "will"

The second common mistake when writing your vision and value statements is to use the word "will" or any other verb that describes the future; rather, it would be beneficial to write "I have" or "I did." Always write your statements in the past or present tense, as if they already exist or have been done. One important reason for this is that it makes it even more difficult to include how and what details in your statements. Another is that it transforms your statements from being a never-ending treadmill of self-sabotage, procrastination, and preemptive scapegoating into powerful mantras that prime you to start living your ideal vision and values today.

Mistake #3. Writing from the mindset of a degenerative or sustainable paradigm

The third common mistake is to phrase your statements in a way that reflects the degenerative and sustainable paradigm. As an example, consider the difference between these three statements:

Working with rather than against nature.
Nature is working for us.
We are nature working.

Although commonly quoted, "Working with rather than against nature" is an example of the sneaky pervasiveness of the sustainable paradigm that believes humans are separate from nature and inferior to it. It's basically saying "We are working with nature, but she is over there." I see the second statement, "Nature is working for you," as firmly based on the degenerative paradigm that also views humans as separate from the natural world, but that it is our slave to do with as we please. You will often see some variation of this statement in the advertisements for biocides or genetic engineering. The final statement, "We are nature working," is elegantly clear that we are part of the natural world and working towards a common goal. It embodies the regenerative paradigm perfectly.

If you think this is just semantics, remember the gorilla illusion and the ability for our unconscious paradigms to shape our perception and thus our principles, practices, and processes.

Mistake #4. Becoming paralyzed by perfection

The final mistake is to become paralyzed by perfection and never finish or even start working on these statements. At the time of writing this line, I have had no less than a dozen "final" versions of my vision and value statements in the last eight years. And each time I finished a one-pager, I experienced a feeling of euphoria reading through the words that I had so painstakingly crafted thinking "There! I finally know what I want!" and "It couldn't possibly get any better than this." Much to my surprise, with each iteration of my vision and value statements, the quality and clarity continue to increase exponentially. I could have never produced my current draft of statements when I started because I did not know myself like I do now, and I am not the same person I was then. With every hour spent in deep reflection, with each conversation with the partners in my life, and with each decision made using these statements, I continue to get clearer about what it is I really want and what is right.

I encourage you to give up waiting for the perfect statements and just get something down that will be useful in the other steps of the process. After all, given the emergent properties of complexity, there is no way you could ever know what your best well-being is. However, if you follow the process, you can trust that you'll continue to hone your values, your vision, and your resources as you design and develop your permaculture property.

> If you follow the process, you can trust that you'll continue to hone your values, your vision, and your resources as you design and develop your permaculture property.

Practices for Step 1: Clarify

Below are three simple but absolutely foundational practices to get you started on your way. We've assumed that you've read this chapter in full, so we've kept the descriptions concise. If you want to download our templates, head to mypermacultureproperty.com.

Practice 1.1. Reflect on What You Have, What You Want, and What Is Right

1. Reflect on the questions:
    - What does well-being mean to you?
    - What do capital, resources, or wealth mean to you?
    - Where do we come from? Why are we here? Where are we going?
    - What are you doing when you feel your best/highest self?
    - What gets you out of bed when times are tough?
    - Describe in detail what you would do if you had one year to live? One month? One week? One day?

TAKOTA'S STORY: Buckets of Well-being

During a video call with a land steward named Mary, the following statement for spiritual resources stood out to me from her vision and values one-pager: "I work hard and carry five-gallon buckets daily."

Apart from its uniqueness, this statement stood out because during my onsite property assessment, her husband, John, had made several comments in frustration about how much time his wife "wasted" carrying buckets around the farm for hours on end despite the fact they had an ATV, skid steer, and several tractors that could help her finish this work in a fraction of the time. I also knew that "increased work efficiency" was one of the key objectives of our consulting contract. Seeing the incongruence here but being conscious of emotional landmines, I genuinely asked: "What is it about this, that is so important to you?" What followed was such a heartfelt and elaborate description that it actually brought tears to my eyes. I was able to draw out words that she had used, and fill in some gaps with some of my own that I thought might fit. Words like strength, usefulness, purpose, productivity, accomplishment, movement, and challenge were all values that resonated with her. Setting an example of a work ethic for her young children and making a contribution to the farm were also key components.

Hard work and carrying buckets was an essential part of Mary's well-being, and judging by her articulate justifications and slightly defensive tone of voice, I was certain this was not the first time she had explained herself before. I could also tell by her body language that Mary wanted this statement to stay on the one-pager and she was prepared to fight for it.

I then asked her the following: "Are there any other activities that provide this same kind of well-being for you in your life?" Mary's whole body softened with the realization that someone was finally trying to understand her, not change her. After some thoughtful reflection, she listed several other examples of hard work on the farm that met her same vision and values as carrying buckets did.

By making disarming statements, I was able to ensure Mary felt secure that my intention was not to challenge her status or autonomy but rather to help her to see that both hard work and carrying five-gallon buckets were not the end goal of well-being itself but simply familiar means that were really about a sense of accomplishment, physical strength, usefulness, children with a work ethic, and a collaborative farm. Mary also had some valuable insights that her existing paradigm had her heading towards a self-fulfilling prophecy of drudgery because, as she later admitted, the hard work was starting to wear on her body. The more her husband tried to help her, the more determined she became to keep doing it to protect her status and autonomy. Spite and martyrdom are really just concentrated forms of violence and silence. If you see someone purposefully making themselves suffer just to prove that they are right, then you can be absolutely certain of two things: they are very afraid something they value is being threatened, and you have a lot of craters to fill in.

For Mary, chores had always been the highlight of her day and because of the way it made her feel. But because she had unconsciously equated hard work and carrying buckets with those feelings for years, she had an unconsciously vested interest in building a permaculture property that was as inefficient and ineffective as possible even though this was the exact opposite to the conscious goal both Mary and John had agreed to for our consulting objectives.

After several sessions, Mary and John came to the realization that they both cared deeply about the same vision and values of accomplishment, physical strength, usefulness, and work ethic, but because they had both mistaken form over function, they ended up creating an impassable no-man's-land between each other and their common vision and values. With two simple questions, they were able to transform a constant point of contention in their partnership into a powerfully motivating goal.

- What are your values?
- Why do ethics matter?
2. In a journal, write down your answers to these questions or have a partner write them down as you speak.
3. Let this journal sit for a week or more, then, with the clarity of distance, go back and read what you wrote.
4. Review anything you wrote down from practice 0.1.
5. Discuss your insights with your accountability partner.
6. Incorporate these insights into your vision and values one-pager (Practice 1.3).

Practice 1.2. Create (and Maintain) Your Personal Resource Inventory

When creating your personal resource inventory, a digital spreadsheet or physical three-ring binder works well, depending on your preference. Both can have tabs for each of the eight resource categories to help you get (and stay) organized. See Figure 1.3 for examples of both. Remember that your personal resources represent the things you want to keep in mind as you are building and managing your permaculture property, and ultimately as you work towards your goals. It'll help you to have a big-picture view of the system, inspire insights about resources not properly leveraged or considered, identify priority and gap areas, as well as help keep you accountable.

1. Go and get a binder with tabs or create your own digital template or download Template 1.2: Resource Inventory.
2. Brainstorm and record your resources within each of the eight categories. See Table 1.1 for common existing inventories that you can leverage.
3. Discuss your insights with your accountability partner.

Further Tips
- Your listed items do not have to necessarily be owned by you, but they are available for your use. For example, if your neighbor has a tractor, and you know that you could easily borrow it, you may choose to include that tractor on your resource inventory. Or at the very least, you may want to list your neighbor as an important relationship within your social resources.

- ▸ Don't be too judgmental or critical of the list when you are getting started.
- ▸ The bigger picture is more important than going into great detail, and your specific circumstances will dictate what level of granularity is important for you to list in your resource inventory.
- ▸ Your personal resource inventory can be a fantastic management tool going forward, and you may find it very useful to keep your personal resource records as up-to-date and reflective of the present moment as possible. In practice though, I have found that this exercise is more about initially setting the stage and inspiring insights. I tend to keep my financial and material resource inventories as accurate as reasonable on an ongoing basis, and the remainder of my inventory only gets evaluated and updated approximately once per year—or as required to make a big decision.

Practice 1.3. Create Your Vision and Values One-pager

Your vision and values one-pager is your goal post. If more than one partner is responsible for decision-making in your project, you'll want to create one document as a team. But it can be helpful to create your individual one-pagers first, before doing so. Try and keep it to one page and watch for the four common mistakes discussed earlier.

1. Download Template 1.3: Vision and Values One-Pager, or create your own based on the sample provided in Figure 1.5.
2. Create a single sentence that articulates *What is right* for you. You may even choose to keep it straightforward and simply go with the three permaculture ethics: earth care, people care, and future care. Put this at the top of the page.
3. Create a single sentence for each of the resource categories that describes the ideal and ethical expression of how that resource will provide you, or support you in achieving, well-being. Make sure to leave space at the bottom so that you can add your vision and values indicators that we will be discussing in Step 5: Monitor.
4. If you get stuck with articulating the right words, do an internet search for "core values word list" or download Template 1.3b:

FIGURE 1.7. Michelle and I invested the time early on to get clear on what we had, and put down, on paper, how we really wanted to live, given our limited numbers of hours on this planet. Getting on the same page and setting a shared vision has been foundational to our well-being and success.

Core Values Word List. Review the lists and choose words that most resonate with you. Next try incorporating them into your statements.

5. Once you've got a first draft, review each statement with your project partners (or better yet, take turns asking each other): *What is it about this that is so important to you?* and *Can you think of any other activities that provide this kind of well-being?* Incorporate any insights into your vision and values statements until you feel you have arrived at the heart of what it is you really want.

6. For each statement, ask yourself: *Is this statement written from a degenerative, sustainable, or regenerative paradigm?* Reflect on the answer and make any changes as necessary.

Credit: Gavin Young

7. For each statement, ask yourself: *If this became reality, would it be good for all people, good for the planet, and good for all conscious creatures into the future?* Reflect on the answer and make any changes as necessary.
8. Reflect on each statement and consider how wide the gap is between your current reality and what you have described. Attempt to order the statements such that the resource category(-ies) with the largest gap(s) are at the top, or towards the top of the page.
9. Discuss any insights with your accountability partner.

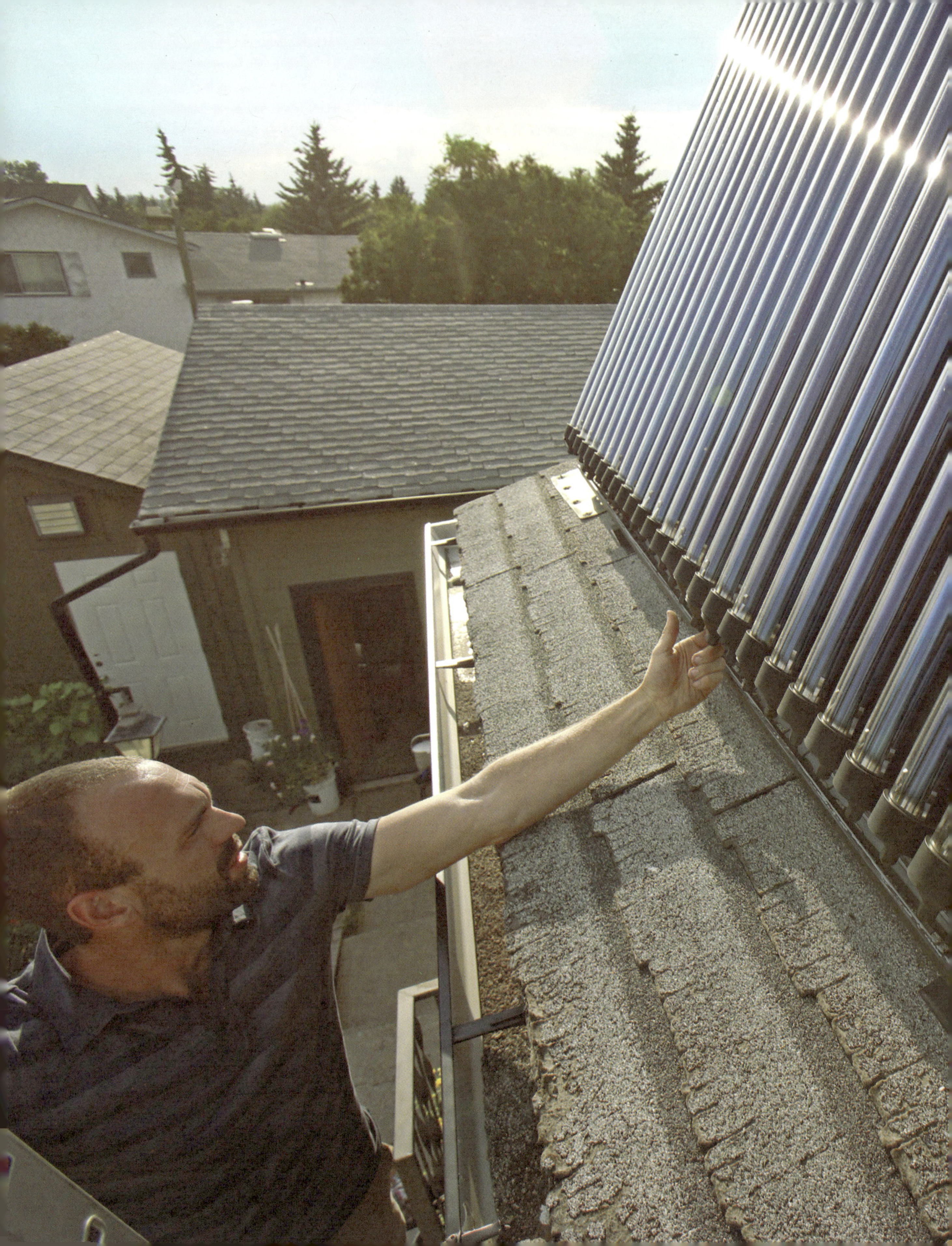

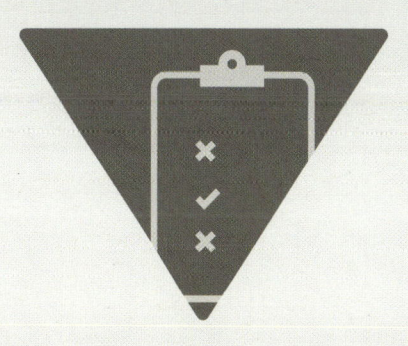

Step 2: Diagnose Your Resources for Strengths, Weaknesses, Opportunities, and Threats

Credit: Gavin Young

Diagnose

di·ag·nose

\ˈdī-ig-ˌnōs

verb

To recognize the cause or nature of a situation by analyzing signs and symptoms.

> *Traditional agriculture was labour intensive, industrial agriculture is energy intensive, and permaculture-designed systems are information and design intensive.*
>
> David Holmgren[1]

The next step in the 5SP Process is about how to overcome the struggle of not knowing where to find information that will aid the design, implementation, and management of your permaculture property. The default approach to solving this dilemma is to consume and attempt to memorize copious amounts of information from many different disciplines, such as farming, biology, ecology, building science, energy systems, geoscience, and soil science. On a scale from cheapest to most expensive, options for gathering information include direct observation, conversations with peers, books, online content (databases, videos, blogs, podcasts, forums), conferences, courses, or perhaps even hiring consultants.

But there are at least three problems with any attempt to memorize information:

1. It is impossible because of the emergent nature of complexity. There will always be more to learn.
2. It is unnecessary because the complexity of life has a several billion-year track record of working just fine without humans influencing or understanding it.
3. It is counterproductive because obsessively consuming information quickly leads to the next struggle of "I don't know how it all

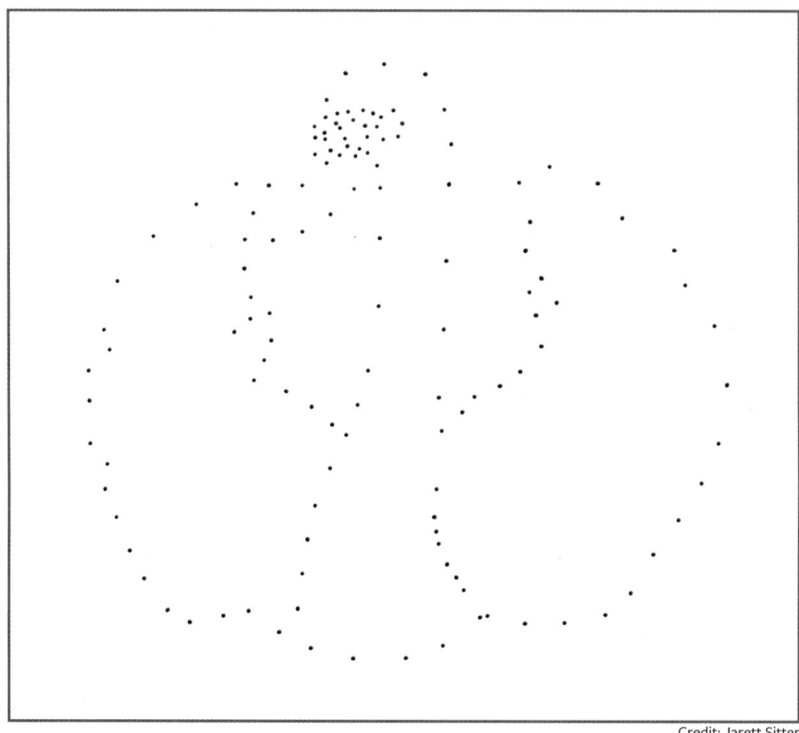

FIGURE 2.1. Can you guess what it is? Turn to page 79 for the solution.

Credit: Jarett Sitter

fits," then on to "I don't know where to start," and finishing with "I don't know when it will end."

Trying to learn everything about permaculture design is like trying to dam a raging river by grabbing it with your hands. Your hands might get wet, but you're not actually storing any water. The solution is not to just have faith that it will all sort itself out either. To overcome the struggle of not having enough information, you need to stop focusing on how to gather and memorize information and start focusing on how to organize it. To illustrate why memorization doesn't work, but organization does, try to solve the connect-the-dot puzzle shown in Figure 2.1, which contains no numbers. Even though the human brain is very good at recognizing patterns, it can only hold and be conscious of about seven pieces of information at one time.[2] With over fifty dots on the page, it's easy to see why you can't tell what the solution is and how quickly your brain enters into the struggle of having too much information.

The reason you cannot solve the puzzle is not because you don't have enough dots; it is because the dots are not numbered. No matter how long you stare at those dots, even if you memorized them, it would not help you find the solution. They need to be organized. The same is true for building a permaculture property, except that a permaculture property has millions of data points. Further, because of the complexity of living systems, some of the dots are missing or erroneous and all of them are constantly on the move! The human brain is built for pattern recognition, not memorization. In contrast, tools like filing cabinets and computers *are* built for memorization—they can store copious amounts of data in files or code. Thus, when we let them do what they do best, we can focus on what we do best. In short, diagnosis is all about organizing data in a way that allows your brain to connect seemingly disparate pieces of information in extraordinary ways to achieve unimaginable results. Turn the page and see how easy it is to solve the same connect-the-dot puzzle now that the dots have been ordered and organized.

To help you start solving your own connect-the-dot puzzle, in this chapter you will learn how to:
- Use a pattern found everywhere in nature to gather, store, and organize any amount or kind of information.
- Describe any pattern on your property, or in the whole universe, in great detail using only four variables.

In short, diagnosis is all about organizing data in a way that allows your brain to connect seemingly disparate pieces of information in extraordinary ways to achieve unimaginable results.

TAKOTA'S STORY: Don't Skip Your Diagnosis!

Imagine limping into your doctor's examination room and telling her, "Hi, Doc, I have patellar tendinitis in my knee. I'm sure of this fact because I read about it online. Save us both some time and just go ahead and book me in for surgery tomorrow please." After a strange look, she'd probably pretend that she didn't hear you and ask you to climb onto the table so that she could begin her own diagnosis. And after a physical examination, she'd ask you a bunch of questions, then refer you to get an ultrasound or an MRI. No matter how persistent you were that you knew exactly your ailment and that you were willing to jump straight into the operating room, she wouldn't skip her own diagnosis (nor should you want her to). What if she booked you in for surgery only to discover on the operating table that cutting you open was not going to solve your problem?

In the same way, when it comes to building your permaculture property, *diagnosis is both the most valuable and the most important part of the whole process, in addition to being the step that many try to skip over altogether*. I can't even count the number of potential clients who, in their initial consultancy, basically stated, "Here's what I need, let's get started with construction tomorrow!" The doctor's examination room analogy is one that I've found myself bringing up most frequently with clients. And for those still insistent on moving fast, I share another common aphorism: *There's cheap, fast, and good, and you can only pick two. Your options are therefore: (1) cheap and fast, but not good, or (2) cheap and good, but not fast, or (3) fast and good, but not cheap.*

Assuming that you want to have a good diagnosis, you have only two options: pay a good consultant to speed up the process or take the time to do it yourself.

Step 2: Diagnose Your Resources   79

- Assess whether a specific pattern on your property is a strength, weakness, opportunity, or threat.
- Summarize your most important diagnosis insights onto a single page that will be used as a checklist for your design.
- Avoid believing in the myth of objectivity.
- Be on the lookout for the one or more pieces of information that can make or break a permaculture property.

A Watershed of Information

Recall that a pattern is a repeating event that results from the interaction of two or more forms of energy or matter. Also, all patterns represent the most optimized expression of that interaction of energy or matter after billions of years of trial and error.

One important and common pattern within nature gets its name from the Ancient Greek word *dendrite*, which means "of or pertaining to a tree." A dendritic pattern, then, is the manifestation of the interaction of energy and matter that takes on the branching form of a tree. Look back to Figure 3, particularly the dendritic patterns that exist throughout nature. A key function of all these patterns is to gather and distribute energy and matter. For example, a river

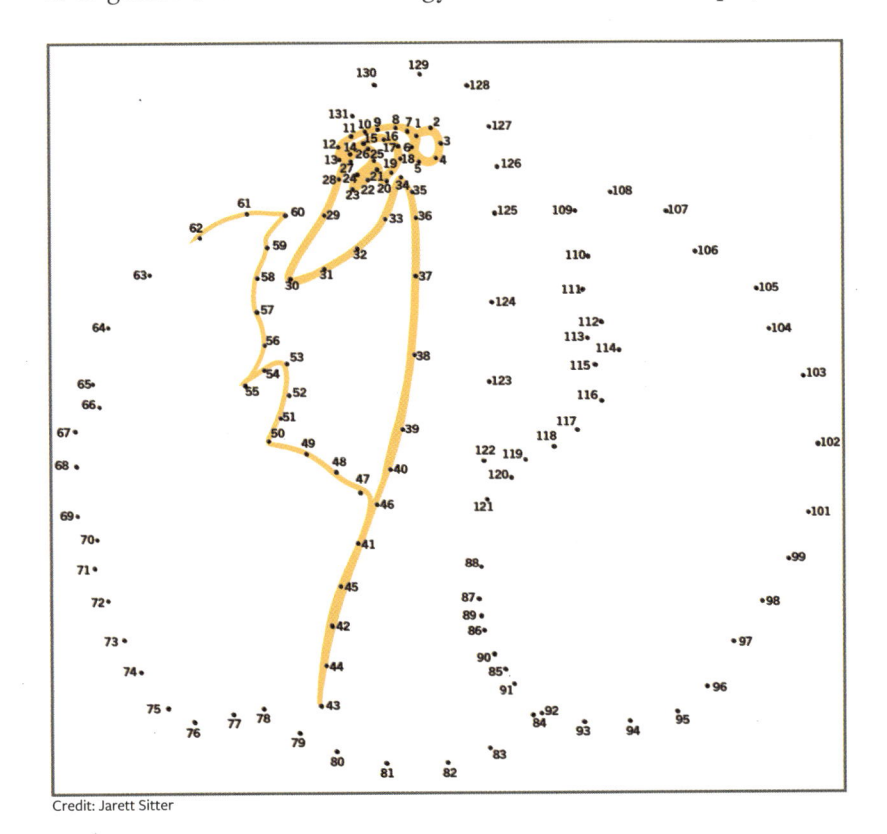

Patterning is the way we frame our designs, the template into which we fit the information, entities, and objects assembled from observation, map overlays, the analytic divination of connections, and the selection of specific materials and technologies. It is this patterning that permits our elements to flow and function in beneficial relationships. The pattern is design, and design is the subject of permaculture.

— Bill Mollison[3]

FIGURE 2.2. Now that the dots are organized and a few connections are made, the solution reveals itself.

Credit: Jarett Sitter

gathers water and nutrients from the land through a dendritic pattern of upland watersheds and then distributes it to the sea via a delta. A delta, also a dendritic pattern, gathers nutrients from the sea via fish and distributes it up the river via spawning cycles to the land. The canopy of a plant gathers sunlight and carbon dioxide and distributes it to its roots, while the roots gather water and nutrients and distribute it to the canopy. An animal's lungs, brains, nerves, liver, kidneys, and veins all gather and distribute oxygen, information, nutrients, and toxins to and from the organs inside our body and to and from the outside world. Mycelium gathers and distributes nutrients to and from soil organisms, soil minerals, soil moisture, plant roots, and other fungi. Communities of humans, like cities, gather and distribute energy in the form of people, goods,

TAKOTA'S STORY | Growing Up a Carpenter

Little boxes on the hillside
Little boxes made of ticky-tacky
Little boxes on the hillside
Little boxes all the same
There's a pink one and a green one
And a blue one and a yellow one
And they're all made out of ticky-tacky
And they all look just the same.

— Malvina Reynolds, singer/songwriter, "Little Boxes," 1962

My first job after high school was working as a carpenter, mainly for my older brothers who both ran their own construction companies. I worked on every aspect of home building from foundations and framing right through to finishing work. We built many custom homes, mansions really, and dozens of cookie-cutter houses like the ones that gave Malvina Reynolds inspiration for her song "Little Boxes."

However, with each new home or renovation I completed, I became increasingly disenchanted with our modern approach to shelter. Malvina's song was quite true, our houses were in fact literally made of ticky-tacky. The paints, polishes, pipes, plywood, putties, siding, shingles, sealants, windows, wallpapers, foams, and fixtures were not only made partially or entirely of plastic, they all came packaged in it. And almost every package or material I installed in these homes had the same ominous label: "WARNING: This product contains chemicals known to the State of California to cause cancer and birth defects or other reproductive harm." When I finally pointed this out to my brother, without hesitating he chuckled, "Well it's a good thing we aren't in California then, isn't it?"

One of my jobsite duties was to load all of the garbage into a trailer and take it to our local landfill where, for a few dollars' tipping fee, it

services, and ideas via its branching roads, power lines, water lines, sewer systems, phone lines, and the internet.

Based on these examples, it is reasonable to see how this pattern could help guide our efforts to gather and store information in our diagnosis process. But dendritic patterns also have a second almost magical function in nature, that is also useful to something else in our 5SP Process. When a city, fungi, animal, plant, or river has gathered enough energy or matter, it gives birth to something entirely new. The city spawns another city, the fungi releases spores, the animal procreates, the plant produces seed, the river plugs and bursts its banks to create a new drainage. In the same way, when you mimic this same dendritic pattern to organize the watershed of information about all your resources, your diagnosis will inevitably

was possible to deliver several thousands of pounds of waste to be buried in a gigantic hole in the ground. This "dump run," as we called it, occurred every couple weeks and numerous times during a project. I was able to see the full scale of the waste stream produced during the building of a modern home. I would estimate that, from start to finish, the amount of waste generated is equal to a pile at least one tenth the size of the home being built.[4] Now when I drive past houses in nearby towns, I know that for every ten houses, there is a pile of carcinogenic, birth defect, and infertility-causing trash buried in a hole in the ground not 15 kilometers (10 miles) from my home.

On my numerous dump runs, I would also notice large semitrailer trucks bearing the names of cities that were located more than several hundred kilometers away. It turns out that the nearby landfill has some of the cheapest tipping fees in the province because of a rare kind of clay upon which it was situated. Because of this clay, no costly plastic liners were deemed necessary to contain the landfill wastes. However, in a chance meeting with an off-duty intoxicated landfill employee, I learned that management was suspicious that leachate was indeed getting into the groundwater. This was one of the last straws that led me to leave the construction industry and come back to the family farm. But while doing the diagnosis of the water resources for my property, I remembered this critical data point, and as you will see later, it is the reason I am developing a water-harvesting and gravity-fed passive irrigation system on my farm that does not rely on the polluted underground aquifers.

give birth to a design. Once again, we see how the compounding, dynamic, emergent problem of complexity becomes an incredible solution.

Your Property Resources

Just like how you organized your personal resources in Step 1 into eight categories (financial, material, living, social, experiential, intellectual, spiritual, and cultural), you are now going to organize your property resources into eleven categories: geography, climate,

TABLE 2.1. Your Property Resources. There are eleven categories of property resources.

Order of Operations	Property Resource Category	Examples/Subcategories
1	Geography	Maps (elevation, hillshade, contours, aspect, slope, greater watershed, property catchment), aerial photos, laws and local regulations, demography, landscape features (ridges, valleys, mountains, lakes, oceans, geology, relief)
2	Climate	Precipitation, temperature (Plant Hardiness Zone, growing degree days), solar, wind, analogue climates, sector map (fire, flood, frost, traffic, view, smell, noise, dust, pollution, etc.)
3	Water	Wells, dams/dugouts/ponds/wetlands, swales, springs, water flows (creeks, seasonal runoff, etc.), diversion drains, culverts, drainage tile
4	Access	Paved roads, gravel roads, farm tracks, foot paths, zone maps, time and motion study
5	Structures	Dwellings, barns, outbuildings, utilities (power, water, internet, phone, natural gas, propane, both buried and above ground, septic field)
6	Fencing	Perimeter, corrals, cross fencing, privacy screens, wind fence, portable fencing
7	Flora	Domesticated, historic and existing plant species. Pasture, gardens, agroforestry, cropland, crop records
8	Fauna	Historic and existing animal species and fungi,[5] grazing patterns, breeding and birth records
9	Business	Historic, current, or future enterprises and their associated accounting, marketing, and administration
10	Technology	Machinery, vehicles, equipment, tools, and other material resources specific to your property
11	Soil	Carbon, nitrogen, phosphorus, potassium, trace minerals, pH, texture, biodiversity, horizons, water storage/infiltration

water, access, structures, fencing, flora, fauna, business, technology, and soil, as described in Table 2.1.

Note that there is some overlap between the personal resource categories in Step 1 and the property resource categories introduced here. While the former are used within a framework to help you build your vision and values one-pager, the latter form the basis particularly of the diagnose, design, and implement steps. I'll discuss the **order of operations** in Step 4: Implement, but in this chapter, let's take the analogy of how both rain and information are gathered and distributed by dendritic patterns a little further.

Harvesting Information with a Filestream

Figure 2.3 shows how a watershed gathers individual drops of water into larger and larger channels starting with raindrops, then sheet flow, before going to rills, runnels, creeks, streams, and then the river, until eventually ending in a delta at the edge of a lake or ocean. The diagnosis and design steps function in the same way. In Step 2: Diagnose, our goal is to create a filestream (pun intended), with the function of converting the torrential downpour of information into appropriately compartmentalized physical or digital file folders. These file folders serve the same function as dams, swales, subsoiling, and gabions to channel and store information into appropriate steam tributaries that correspond to the eleven property resources.

Once these eleven tributaries begin filling with information, they will inevitably flow downstream and deposit themselves as design ideas that start with broad brushstrokes down to minute details, or what David Holmgren refers to as "design from patterns to details."[6] The way that I like to think about this analogy is that the river is diagnosis and the delta is design (more on design in the next chapter).

As soon as I started to mimic the dendritic branching pattern of a watershed to gather and distribute information, my obsession about learning everything there was to know about permaculture vanished. I stopped bingeing on information memorization. This was because I started to notice that the amount of information falling into the catchment of my mind, just like the amount of precipitation falling in a watershed, can exceed the capacity of the filestream or water channels. And when this happened, it was inevitable that the flow of data or rain would burst its banks and flood onto my physical

The river is diagnosis and the delta is design.

84　　Building Your Permaculture Property

or digital desktop as unfiled resources. In other words, information, no matter how good the quality, is only as good as your ability to put it to productive use.

That being said, you don't want a drought of information either, because the speed and quality of your design insights are directly proportional to the quantity and quality of your data; poor information yields poor design. You don't want torrential downpours of data; you want a slow and steady drizzle that keeps pace with the evolution of your filestream. You will also find that just as an older watershed that contains high-carbon soils and deep-rooted vegetation can handle more rain and even the occasional flash flood, a more established filestream can better slow, spread, and sink the occasional higher flows of information.

Figure 2.4 and Figure 2.5 show two examples of digital filestreams that I and my co-authors have used. Note that nested subcategories each contain information that increases in detail and quantity. In addition, where the nested subcategories represent the creeks, runnels, rills, sheet flow, and rain, you should organize your subcategories (and subfolders) such that you don't create more than four or five nested layers. Not only is this an observable pattern from rivers and trees and dendrites found in nature, I can attest from experience that more than four nested subfolders/layers layers becomes difficult to manage (see Figure 2.5 for an example).

FIGURE 2.3. (*opposite*) The eleven property categories (geography, climate, water, access, structures, fencing, flora, fauna, business, technology, and soil) can also be organized in a dendritic pattern, similar to a watershed. The categories are simply branches of the river of resources. Each category further branches into nested subcategories with information increasing in detail and quality that can be described using the four variables of pattern (form, timing, placement, and scale).

FIGURE 2.4. In Google Earth Pro, which we will discuss more later, your filestream can be set up, or downloaded from our website, and appears in the Places panel, under My Places. The image here shows the filestream that Takota uses in his own farm design.

Credit: Takota Coen

86   Building Your Permaculture Property

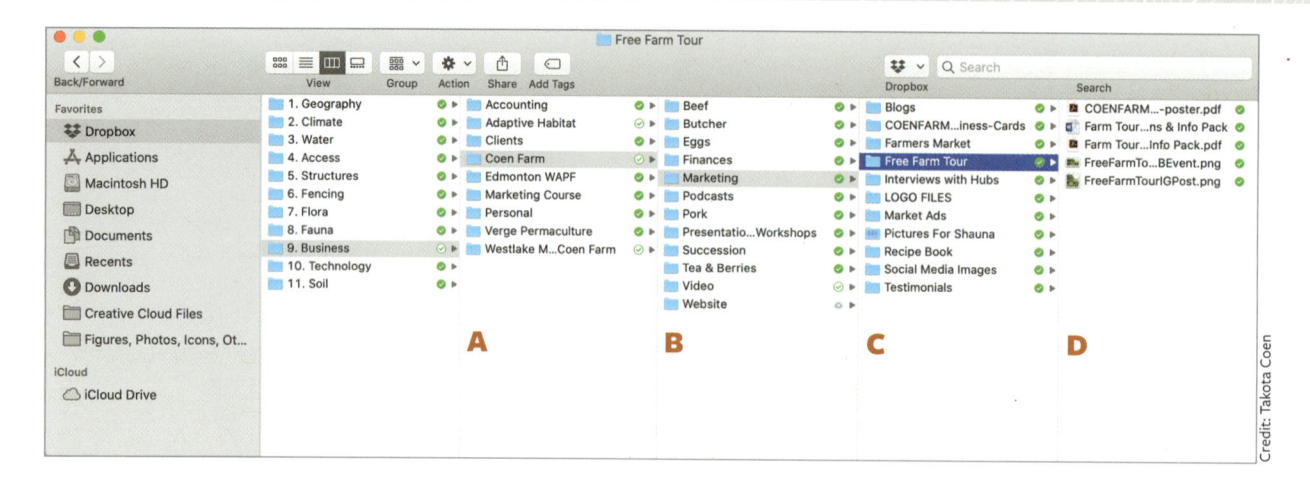

FIGURE 2.5. Computer file folders organized using the 5SP Process filestream. The business category is toggled open showing four nested subfolders/layers: (A) Coen Farm, (B) Marketing, (C) Free Farm Tour, (D) Specific Files. Note that more than four nested subfolders/layers is not recommended.

I encourage you to take some time to familiarize yourself with the more detailed subcategories of the filestream that are available on our website, mypermacultureproperty.com. As you start to experiment with where information should go in the filestream categories and subcategories, you will be able to use your own reason, even preference, to modify the existing pattern to suit your needs. For example, many of the subcategories can be placed in multiple locations. Your files with information about your buried water lines could go under "Water," or with the rest of your utilities under "Structures." Do whatever is most intuitive for you. You may also find the need to add subcategories that I did not include. Alternatively, you may want to rename one of the categories. The only change I encourage you not to make is to delete one of the eleven main categories, even if you don't have an immediate need for it. Trust me in that, if you use this process long enough, there will come a time when all of these are necessary.

As you start to use and adapt this filestream to organize your existing and new information about your resources, remember that it has two purposes:

1. To function as an infinitely expandable, highly intuitive, organized, and hierarchical system to gather information about your property resources.
2. To create a sensible and ordered workflow, called the order of operations, for both design and implementation. This functionality will be covered in greater detail, particularly in Step 4.

Also, look back to Figure 2.4 and Figure 2.5 and notice that the digital folders are numbered so that they appear or can be organized in the correct hierarchy (instead of defaulting to an alphabetically ordered list). If you are old-school, like me, this same filestream structure can be used to organize your filing cabinet. It may take you a day to rearrange all your existing physical and digital information to this new format, but it is more than worth the time. Once it becomes a habit, you will find that struggle of where to find information will disappear and become as natural and effortless as water running down a hill.

It's worth reminding you that not only do the property resources have a hierarchy but you also intentionally ordered your vision statements on your one-pager (see practice 1.3, #8). While the filestream order is fixed, the order of your vision statements is completely unique to your situation, subjective to your own rating system, and, as you will see in Step 4, may even change through time.

Two Stages of Diagnosis

Now that we have a dendritic pattern to mimic that will help us gather and organize information, it is time to start putting new dots on our piece of paper. Soon we will start diagnosing your personal and property resources for strengths, weaknesses, opportunities, and threats, but making value-based judgements (e.g., saying something is a strength or a weakness) is actually the second stage of diagnosis. In the first stage, you must resist the temptation to start labelling things as good or bad right off the bat. You need to simply collect observations about your personal and property resources that are as unbiased as possible. To do this, I recommend using something that I call the four variables of pattern.

Stage 1: Use the Four Variables of Pattern to Make Observations About Your Resources

Several years ago, I was trying to come up with an analogy for the iterative and often irritating feeling of designing and developing a permaculture property. What first thing came to mind was an old wooden ball maze I used to play with as a child. The game required the turning of two knobs that tilted a platform with the goal of navigating a steel ball along a narrow path littered with holes. If you hit a hole, the ball would go back to the start and draw forth streams of profanity from me. It resonated immediately. But with this metaphor in mind, I began to ask myself what were the knobs that I was manipulating on my permaculture property to navigate

> While the filestream order is fixed, the order of your vision statements is completely unique to your situation, subjective to your own rating system, and may even change through time.

myself through the maze of suffering and towards the finish line of well-being?

I shared the analogy with my co-authors, and after many discussions, we agreed that in permaculture design, the knobs that controlled the playing surface were form, timing, placement, and scale. Or in other words, building a permaculture property is a never-ending maze of trying to do the right thing at the right time in the right place in the right amount. These four variables and their rela-

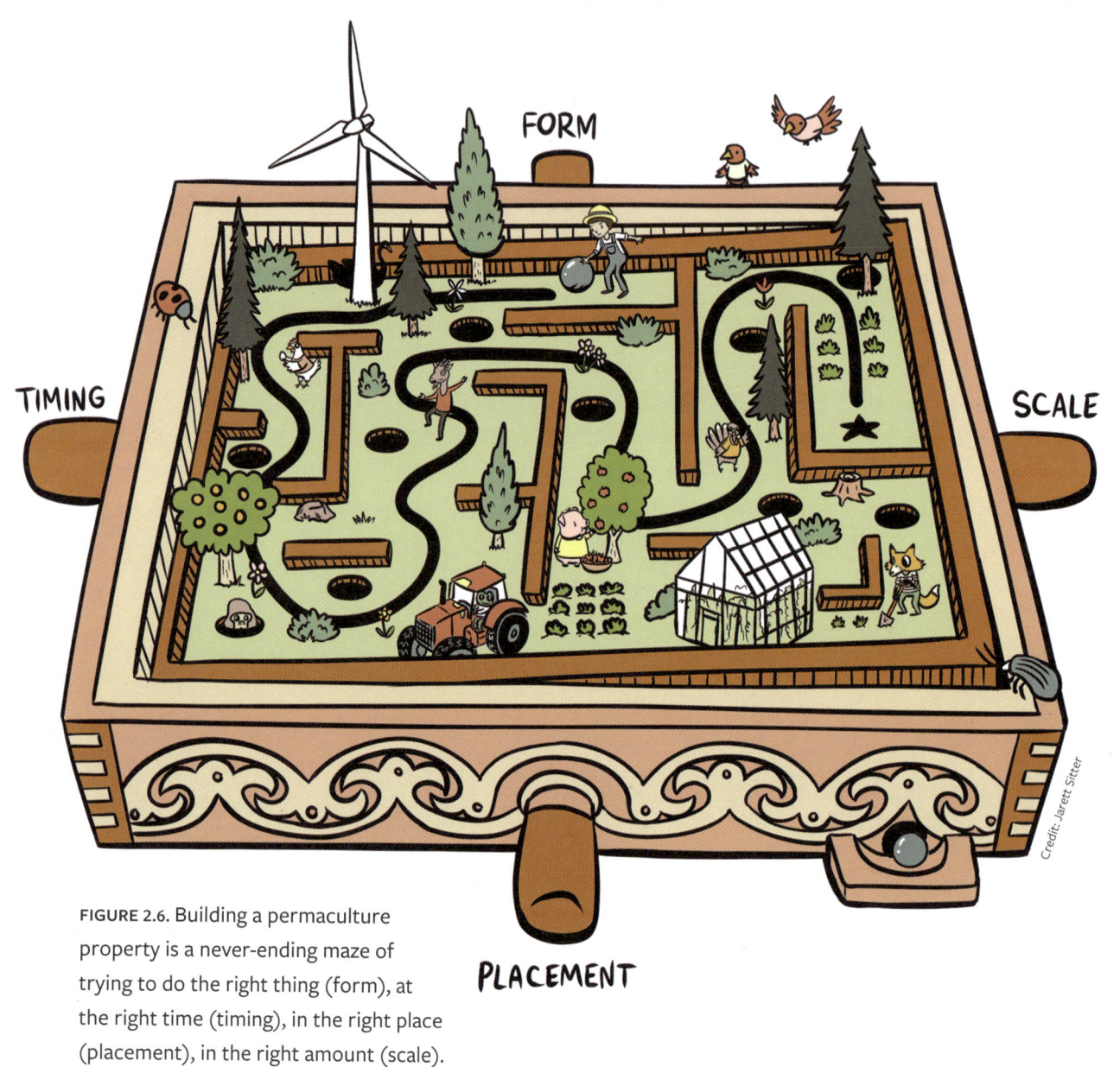

FIGURE 2.6. Building a permaculture property is a never-ending maze of trying to do the right thing (form), at the right time (timing), in the right place (placement), in the right amount (scale).

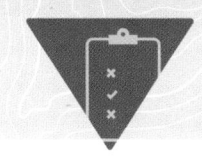

tionship with design, implementation, and management is a theme that will continue through the entirety of the 5SP Process. In this chapter, we will focus on using form, timing, placement, and scale to simply describe your property resources or really, any pattern in the universe.

Looking at the potential subcategories for your property's resources shown in Table 2.1, it is easy to get overwhelmed by the task of collecting information and head into the other struggles of permaculture, how does it all fit, where should I start, and when does it end. My recommendation is to simply consider each relevant subcategory using the four variables of pattern: form, timing, placement, and scale. For example, when diagnosing the precipitation patterns of climate (the second property resource category), you would look at the annual average, maximum, and minimum of the various forms of precipitation and the intensity of single events—this is scale. The duration, frequency, and time of year various precipitation events occur—this is timing. The prevailing direction weather systems come from or locations where snow might concentrate—this is placement. The kind of precipitation you get like snow, rain, hail, and condensation—this is form. This same detailed description would be repeated for your solar, temperature, and any other sector patterns associated with climate and for each of the other resource categories and the associated subcategories relevant to your particular property. See Table 2.2 for a list of other words that are common when making observations and how they fit into each of the four variables of pattern. You can start to see now why having a good organization system is essential for all this data, and any attempts to memorize it would be futile.

In this chapter, we will focus on using form, timing, placement, and scale to simply describe your property resources or really, any pattern in the universe.

TABLE 2.2. The Four Variables of Pattern. The four variables of form, timing, placement, and scale can be used to diagnose and gather information about any pattern.

Variable	Ways to describe
Form	Shape, color, relative quality, morphology, intrinsic characteristics
Timing	Frequency, duration, cycle time, schedule
Placement	Spatial location, orientation, direction, connections, relationships, angle
Scale	Size, weight, intensity, quantity, amount, strength, efficiency, depth, length, width, girth, distance, speed, area, concentration

The first stage of diagnosis is simply about describing the four variables of pattern for your eleven property resources, the second stage is about making value judgments about whether these patterns are helpful or harmful to your well-being clarified in Step 1. Returning to our metaphor of the dendritic pattern of information, you can see that Figure 2.3 shows how the last order or smallest branch of each tributary is none other than the form, timing, place-

The Myth of Objectivity

In the Six Ps of Epistemology model introduced at the start of this book, I mentioned that there must be, at some level, objective patterns in the universe. Presently those seem to be things like water runs downhill, hot air rises, cold air sinks. However, I also mentioned that, because of our paradigms, it is impossible that humans can be completely objective about those facts. Let me substantiate this by first diving into the concepts of precision and accuracy. Where precision is how finely measured an observation is, being one hundred percent precise would mean that the measurement can be exactly replicated over and over again. Where accuracy is how close your observation is to the truth, one hundred percent accurate would mean that your observation is unfalsifiable, it is truly the way things are, or it is absolutely correct. The concepts of precision and accuracy are illustrated using the analogy of a bulls-eye target in Figure 2.7. To be one hundred percent objective, you would need to be one hundred percent precise and one hundred percent accurate.

However, it is first impossible to be one hundred percent precise for at least three reasons:

1. The outcome changes depending on how you measure the pattern. For example, when measuring the coastline of any land mass, a more detailed measurement does not result in an increase in accuracy—the measurement only increases

FIGURE 2.7. Precision versus accuracy.

Credit: Jarett Sitter

ment, and scale. You can then imagine patterns as the rain that falls on the landscape. That rain can then be diverted into form, timing, placement, and scale and then conveyed by swales into the various dams (folders) that exist in progressively larger channels of your filestream.

As shown in Figure 2.4, the Google Earth Pro filestream Takota uses for his own farm design (more on that later), you can see the fol-

---

in length. There is no way to obtain a maximum value for the length of a coastline (known as the coastline paradox).

2. The act of measuring it changes the outcome of the pattern. For example, limnologists (those who study lakes) can be an important factor in the spread of lake organisms that they study via their boats, boots, and nets.
3. The pattern itself changes before you have finished measuring. For example, even if both the cartographer and the limnographer could complete a 100% precise survey of the coastline or the critters by the time they discovered their objective total, the value would have already changed because of changing water levels or population growth.

Next, it is impossible to be one hundred percent accurate because our assessments are clouded by our paradigms. As Sam Harris puts it: "the very idea of 'objective' knowledge (i.e., knowledge acquired through honest observation and reasoning) has values built into it, as every effort we make to discuss facts depends on principles we must first value (e.g., logical consistency, reliance on evidence, parsimony, etc.)."[7] The biggest issue I see here is a real proclivity on the part of humans to claim that their observations are indeed accurate, yet these are steeped in biases stemming from the degenerative or sustainable paradigms. In other words, just because you do not like something does not mean that something should not exist. Let me give you a common example of this. Labelling something as a "pest," or as a "weed," or as an "invasive species" is not an accurate observation, nor is it objective. Period. But our cultural acceptance that some species are inherently "bad" and our need to dominate or control ecosystems has led to an understanding that justifies the application of billions of dollars worth of toxic chemicals into the environment annually[8] that are undoubtedly contributing to the needless suffering of many organisms, humans being one of them.

You must accept that it is impossible for you, and others around you, to be one hundred percent precise and one hundred percent accurate. And this is why starting this entire process with an inspection of your own paradigm is so crucial. As you cannot help but be subjective in all your observations and actions, the best thing you can do is to consciously base your subjectivity on a foundation of ethical well-being. When you make observations in the first stage of diagnosis, be humble, be childlike in your inquiry, and admit that you do not, and cannot comprehend everything. Remember, you don't need to understand exactly how the sun works (or to label the sun as good or bad) to later benefit from its warmth.

lowing subcategories for Access: paved and gravel roads, foot paths, farm tracks, zone maps, and time and motion study. For all of these, Takota would have drawn or collected information about where they are located (placement), what they are made of (form), when they were built and frequency of use (timing), and their lengths, width, catchment area, and gradient (scale). Note that you may not need all variables to describe every pattern in your resource categories. These four variables (scale, timing, placement, and form) will help you to observe and describe any pattern in the universe, or at least how it appears to be, as thoroughly as possible without any value-based judgments attached. For example, "I have $400,000 owed to the bank on my mortgage, with an amortization period of twenty years and my monthly payment is $2,000" and "There is a wildlife corridor running from the northeast corner of the property to the southeast corner that hundreds of elk transit at dawn and dusk" are suitable statements without value-based judgments attached. In contrast, a statement at this point about the debt or the elk being bad are value-based judgments, not observations.

Stage 2: Performing a SWOT... (sss-What?)

In this second stage of diagnosis, you will now make value-based judgements about your personal and property resources. My preferred tool for making these value judgments is known as a SWOT analysis, an acronym for strengths, weaknesses, opportunities, and threats. In this exercise, you identify things that are helpful or harmful to achieving your vision. Strengths and weaknesses are those attributes that are existing for you or your property. Opportunities or threats are deemed probable future attributes of the neighbors, community, and/or bioregion. Table 2.3 and Table 2.4 show you examples of possible strengths, weaknesses, opportunities, and threats for both property resources and personal resources.

Building on the observations from stage 1 of diagnosis, you could now make value judgments like: "My financial debt is a weakness that is going to limit my ability to purchase new material capital" and "The elk are a threat because they might eat my fruit trees."

However, you must be cautious that your labelling of a pattern or calling a resource good or bad doesn't restrict your ability to turn the problem into the solution. Perhaps you could turn your mortgage expense into an investment opportunity for family and friends? Could those marauding elk become a source of manure, meat, or

> **You must be cautious that your labelling of a pattern or calling a resource good or bad doesn't restrict your ability to turn the problem into the solution.**

fire management? The goal of the SWOT analysis then is not only to analyze your resources but also to analyze your own paradigms and help you get out of your own way on your path to well-being. Separating your observations from your value-based judgments makes it much easier to transform weaknesses and threats into strengths and opportunities, simply by shifting your perspective. Of course, there will be some value judgments, like living next to a toxic landfill, that are hard to see optimistically.

Once you have completed a high-level SWOT summary of all your observations for each of your personal and property resources, you will find that you are already creeping into the design phase. You will also find this SWOT summary very useful when it comes time to decide where to start or what's next in the 5SP Process when we get into implementation.

Black Swans

If nothing in this step has enticed you to put in the hundreds of hours required to do a proper diagnosis of your resources then perhaps this will.

From experience, I have found that every single property has hidden, somewhere in its various resources, at least one piece of information that, if discovered, changes everything for the better. And if not discovered, for the worse. I call these pieces of information "Black Swans" in tribute to a book of the same name by Nassim Taleb.[9] In Taleb's own words, a Black Swan is "an event with the following three attributes. First, it is an outlier, as it lies outside the realm of regular expectations, because nothing in the past can convincingly point to its possibility. Second, it carries an extreme 'impact.' Third, in spite of its outlier status, human nature makes us concoct explanations for its occurrence after the fact, making it explainable and predictable."[10] While the Black Swans you will be searching for on your property will not change the face of the world like the internet, World War I, or 9/11, the discovery that your lower pasture rests on millions of dollars worth of gravel, or that you have a 100-year-old apple orchard hidden in the forest behind your house, or that you live on an active **alluvial fan** or a fire-prone forest that is twenty years overdue its periodic burn, can change the entire scope of the design and development of your property and even save your life. And yes, these are all real examples of Black Swans that I have discovered for past clients during the diagnosis phase.

> **Every single property has hidden, somewhere in its various resources, at least one piece of information that, if discovered, changes everything for the better. And if not discovered, for the worse.**

TABLE 2.3. Property Resources SWOT. Examples of some possible strengths, weaknesses, opportunities, and threats for each of the eleven property resources.

Property Resource Category	Strengths	Weaknesses	Opportunities	Threats
Geography	Good topography, slope orientation, view, cold air drainage, subsoils	Steep land, bad orientation, boggy, solonetzic subsoil, contaminated subsoils	Lenient regulations, helpful neighbors, active community, flexible landlord	Strict regulations, difficult neighbors, isolated community
Climate	Diverse range of good microclimates, wind shelter, growing zone, high number of frost-free days, pleasant range of weather patterns that bring both rain and sunshine	Abundance of harsh microclimates, lack of wind protection, unsuitable growing zone, low number of frost-free days, rain shadow	Grazing your neighbor's weather-damaged crops, better adapting business to climate	Hail, chinooks, tornadoes, frost, smog, fires, drought, storms, noise, dust, pesticide drift
Water	Rain, creek, spring, well, irrigation, ponds, tanks, automatic livestock drinkers, gravity water	Salty groundwater, no catchment, poor quality and quantity, not suitable for livestock	Easy access to water rights to creek, seasonal runoff to be captured in dam/dugout	Flood plain, drought, expensive water rights, spring drying up
Access	Pavement, extensive farm tracks, animal laneways, yard design, good parking	No property security, no space to turn equipment around, no/poor existing farm tracks	Highway frontage for market booth, farmers are open to workshops on farm markets	High local theft, snow drifting, road bans/restrictions, long distance to pavement
Structures	Storage buildings, power, utilities, dwellings, heated shop, barns, well maintained	No or poor storage, or livestock structures, need repair immediately, no dwelling	Repurposing existing structures, renting neighbor's structures, good local trades	No local trades, restrictive building regulations, expensive local materials, fires

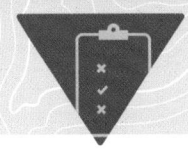

TABLE 2.3. (cont'd.) Property Resources SWOT. Examples of some possible strengths, weaknesses, opportunities, and threats for each of the eleven property resources.

Property Resource Category	Examples of... Strengths	Weaknesses	Opportunities	Threats
Fences	Suitable fencing, well-built and maintained	Poor fencing, need repair immediately, no corrals/handling facilities	Cost-sharing fences with neighbor, repurposing existing fencing, grants for fencing	No predator/pest protection fencing, poor fences on neighbor's land
Flora	Existing genetics, good diversity, good historical yields	Poor existing genetics, crop disease/pests, noxious plants, no yield history	Heritage seed suppliers, increase pasture diversity, local food/crop wastes	Plant pests and diseases from neighbor's monoculture, overspray from industrial farmland
Fauna	Existing genetics, existing predator/pest protection, good diversity	Poor existing genetics, crop disease/pests, noxious plants, no yield history	Custom grazing neighbor's livestock integrating new complementary livestock species	Predators very active in the area, pests and disease spreading, local feedlot
Business	Complementary business to existing model, direct marketing existing farm products	Overlapping business models with existing farmers, land is not certified organic	Existing customers, no competitors, potential to expand, local mentors	Competitors, distance to market, no customer base, saturated market, regulations
Technology	Good existing equipment, sharing capital expenses with existing farmers	No equipment and capital, no cell service, no internet	Renting neighbor's equipment, local rental shop, local farm store, grants	Power/utility supply won't allow future expansion, no local farm stores
Soil	Excellent soil, on farm composting program, existing livestock for manure for gardens/crops	Poor soil health, erosion, compaction, bare soil, pesticide residues, rocky soil	Free compost/manure from community/neighbor	Future salinization problems from irrigation, pesticide runoff from neighbor

TABLE 2.4. Personal Resources SWOT. Examples of some possible strengths, weaknesses, opportunities, and threats for each of the eight personal resources.

Personal Resource Category	Examples of…			
	Strengths	Weaknesses	Opportunities	Threats
Financial	Have accumulated/saved enough money to put a down payment on a property and buy some assets.	Do not have enough money to buy all of the assets and tools needed to make my dreams come true.	Can ask my neighbor to hay my fields with his tractor for a small fee/percentage of yield and sell to market as clean hay/straw. Can sell produce to a market, and once established enough, run workshops for a fee.	Will be in debt in order to cover my asset/tool purchases.
Material	Aside from some key things, I have all the things I need to get started.	Far too many things that are unused and unnecessarily attached too.	Inventory everything and sort into sell, donate, recycle, or keep in order to make more space and financial capital.	Potential to become completely overwhelmed with storing unused things, and inhibits use of productive space for more regenerative purposes.
Social	Make friends easily, adaptable in a wide range of social situations; a lot of friends to ask for help when needed	Friends are far away and tend to not visit.	Forge new friendships and build new guilds with them whereby we all learn from each other and share tools.	Will be a transition period that may be hard between moving to new property and could result in lost productivity.
Living	Health is fairly in order with no major life-threatening ailments.	At times have inflammation in the joints that makes work harder than other times.	Chance to grow nutrient-dense food that nourishes/regenerates my body and do exercises that facilitate strength and antifragility in these regions.	Sometimes the pain is bad enough that work is impossible and nothing will get done.

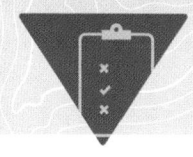

TABLE 2.4. (cont'd.) Personal Resources SWOT. Examples of some possible strengths, weaknesses, opportunities, and threats for each of the eight personal resources.

Personal Resource Category	Examples of... Strengths	Weaknesses	Opportunities	Threats
Experiential	I have strong research and analytical skills as per many years of pursuing what I believe to be the right way to live.	Have limited gardening and farming experience whatsoever.	Book tours with farmers around me (will also help with social capital), reach out to experts, read books in the evening when work is done.	Possibility of overwhelm due to lack of experience and practical knowledge. Furthermore, could potentially make expensive mistakes, or worse, harm myself.
Intellectual	Have a great deal of intellectual property that helps me to generate passive income.	No experience writing, speaking, or sharing my ideas in any way.	Begin writing and producing content as a means to track my own learning process as I build my permaculture property.	I have a contract with my current employer locking up all of my inventions and ideas.
Spiritual	Open-minded and experience with meditation, art-making, journaling, and other outward forms of expression. Have an incredible drive to change the world around me and make it better for future generations.	Outward forms of expression (art, journaling) and inward reflection (meditation) tend to take a place on the back burner when things get busy and there is a lot to do.	Establish a routine for art-making, journaling, and meditation.	If frequently ignored and not addressed, mental well-being and creativity diminishes, and suffering ensues.
Cultural	Am well travelled and have visited with many regeneratively minded folk, have a strong sense of community, and cherish the stories that people bring to the table.	I have no current connection to my surrounding community.	Host or attend potlucks, research local community groups that you would consider joining.	A lack of community cohesion can make me feel that I am working all alone and am not sharing my rewards with anyone/for anything, and suffering ensues.

TAKOTA'S STORY: Black Swan Dam

One of the biggest Black Swans that I discovered on my farm embarrassingly recently was an excellent dam location. In fact, it is the largest and most cost-effective potential water storage on my property. With an estimated cost of $5,000 to build, this 60-meter-long (200-foot), 2-meter-high (6-foot) dam wall would create a pond that could cover an area of 2 hectares (5 acres) and hold back more than 7.5 million liters (2 million gallons) of water. The reason I am so embarrassed to admit this discovery is that five years earlier I built a $20,000 dugout that only holds half as much water, a stone's throw away. Ironically this dugout was built in an attempt to solve another Black Swan, which was the discovery that our aquifer was declining in quality and quantity. The only well on the property that accessed this groundwater would not keep up with new plans to irrigate fruit trees, and could potentially fail altogether, leaving the farm without water. Not to mention the potential sources of pollution from our nearby landfill.

My original property design, which I drafted up many years prior, was done with a single laminated topographical map, erasable markers, vellum paper, and lots of photographs to capture all the information that I had gathered about my farm from decades of my personal observations and my family's history on the land. But having grown up on the land and feeling that I knew it like the back of my hand, and tiring of the difficulty of designing on paper, I started doing most of it in my head. Big mistake.

After discovering that GIS software (specifically Google Earth Pro) would allow me create, organize, and layer geographical information, as well as provide the ability to add lines, shapes, location pins, georeferenced photos, and text effortlessly, I started using digital mapping tools exclusively in my subsequent design, consulting, and education work (more on this in Step 3: Design). Over time, and through many co-consultancies and property design workshops co-hosted with Verge Permaculture, the 5SP Process also started to take shape.

As such, somewhat recently, I finally found a free day to rediagnose my own property and look at it with fresh eyes, this time using the framework and tools that I had been advocating to others. I didn't want to be a hypocrite after all! Starting of course with geography, I was completing the exercise of mapping out all the ridges and valleys on the landscape when the new potential dam site jumped off the screen and slapped me in the face. This incredible Black Swan had been right under my nose for years, I had walked past it dozens of times a year, but because it was covered by a few trees, I missed it.

This was a humbling lesson of not only the importance of diagnosis but the value of ongoing re-diagnosis. I have yet to build this dam that will allow me to do things like wild rice production, ram pumps, fish rearing, and flood irrigation, in addition to increasing my water resiliency, but when I do, I already know what I am going to call it: Black Swan Dam.

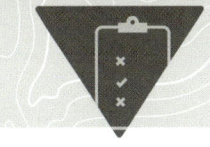

The Value of Digital Mapping and Open Data

Software development and the internet in the last decade have enabled the ability to visualize, create, access, and, particularly, *organize* information in ways that have had profound and positive implications for the permaculture designer.

In some of my early property consulting engagements, in around 2010, I would easily spend upwards of $5,000 acquiring satellite imagery, or hiring a surveyor or a drone operator to create topographical maps. These costly maps would then be printed out as large as possible at the local print shop and pinned down on my drafting desk. Layers of translucent paper, colored pencils, stencils, an architect's scale ruler, and aerial imagery would be used to add scaled map overlays, ideas, any diagnosis, as well as to create designs. I'd buy packs of erasers in bulk. Despite trying several different computer-aided design programs over the years in an attempt to alleviate the rework that came with every change in pencil line, to be honest, up until 2015, I was still doing much diagnosis and design work by hand, then sending it to a draftsperson to make it look pretty for my client.

On top of that, using and accessing data through much of the last decade has also not been easy. Even in 2018, if I needed information, such as historical temperatures, I'd end up on some government website where I would find instructions to install a download utility along with programming command lines. Having no idea what any of this meant, it would inevitably result in a phone call to my brother-in-law (who is a computer whiz). He'd pull the data for me, which would arrive in a very large file with a long list of numbers, and then I'd need to ask Michelle (a spreadsheet guru) to get it in a format that I could visualize and utilize. As such, anyone who didn't have easy access to a computer whiz and a spreadsheet guru could be out of luck.

What's really astounding to consider is how fast all of this has changed in only a few years, and how it is likely to continue to rapidly evolve. Notably, in 2015 Google decided to make Google Earth Pro (GEP) free to the public. This digital mapping tool, called a **Geographic Information System (GIS)**, is a software program that allows you to create, organize, and visualize information based on a three-dimensional globe of the Earth. The ability to zoom in and out to view your property at any scale allows you to effortlessly see

> **Software development and the internet in the last decade have enabled the ability to visualize, create, access, and, particularly, *organize* information in ways that have had profound and positive implications for the permaculture designer.**

the big picture or simply focus in on the minutia, bringing your diagnosis and design skills to a whole new level. Not only does GEP come with a rich database of current (and historic) aerial imagery, placemarks, road names, weather, demographic information, and more, you also have the ability to add and remove colored lines or polygons, georeferenced images, information, Global Positioning System (GPS) coordinates, data, topographical maps, climate maps, or pretty much any geographic information that you can get your hands on, as long as it is in the correct format for import.

Also fundamental to GIS is the ability to create separate layers and ultimately a stack of layered information about a geographic area (i.e., your property), as illustrated in Figure 2.8. Each layer can be separately toggled on and off, similar to how you might use layers of translucent paper in a pen-and-paper version, but infinitely more expandable and versatile.

Another huge advantage of using GIS software is the ability to then download a GPS app for your mobile device or smartphone (or alternatively get yourself a handheld GPS device) and then use your smartphone to create georeferenced photos, text files, **waypoints**, tracks, or simple coordinates that can be imported directly into the GIS software as layer. For instance, I used a smartphone to trace over eight kilometers (5 miles) of hiking trails, along with pertinent photos at certain locations, and imported these tracks and photos into my GEP file. I can now toggle these access trail layers and photographic cues on and off as I am considering design options for my property.

Although I certainly expect the specific tools to evolve (it is a complex system after all), and one day Google Earth Pro may not be our software program of choice, the general idea is this: more and more people are contributing to more and more

FIGURE 2.8. Geographic Information Systems (GIS) are based on layers of information. The filestream not only provides the workflow (i.e., where to start and where to end), it also sets the order of the layers (and sublayers) within the stack of geographical information.

Credit: Jarett Sitter

software combined with more and more **open data**. I believe that digital mapping and GIS software provide enormous value during this time of rapid expansion of data availability and ever-increasing collaboration.

Practices for Step 2: Diagnose

Similar to Step 1: Clarify, there are three recommended practices for Step 2: Diagnose. Much of the nuance in completing these practices is discussed in the preceding chapter, so I don't recommend that you dive in without reading the chapter in full. Lastly, if you want to download our templates, head to mypermacultureproperty.com.

Practice 2.1: Organize Your Existing Information

Whatever your preferred medium, digital or analogue format, start by organizing any of your existing information systems into a dendritic filestream so that you can get, and stay, organized. This could include your physical filing cabinet, your computer folder directory, and perhaps, like me, you may even choose to organize your library books and photo albums in this same way. Just remember to use a storage format that is easy to rearrange and navigate, and can branch out into infinite subcategories. Look back at Figure 2.4 and Figure 2.5 for examples of how to organize information using the 5SP Process filestream and Table 2.1 that lists the eleven property categories, some possible subcategories, and their hierarchical order. You can create your own templates based on these, or alternatively download one of the following:

- ▸ Template 2.1: Computer Folder Filestream
- ▸ 5SP Process Filestream for Google Earth Pro

Practice 2.2: Gather New Information

Now that you have started to retrain your brain and your filing system to organize information like a watershed, you can begin to gather new detailed information about the form, timing, placement, and scale of each of your property resource categories and store them in their appropriate folders in your digital, GIS-based, or physical filestream. Start with geography and work your way down

> **Digital mapping and GIS software provide enormous value during this time of rapid expansion of data availability and ever-increasing collaboration.**

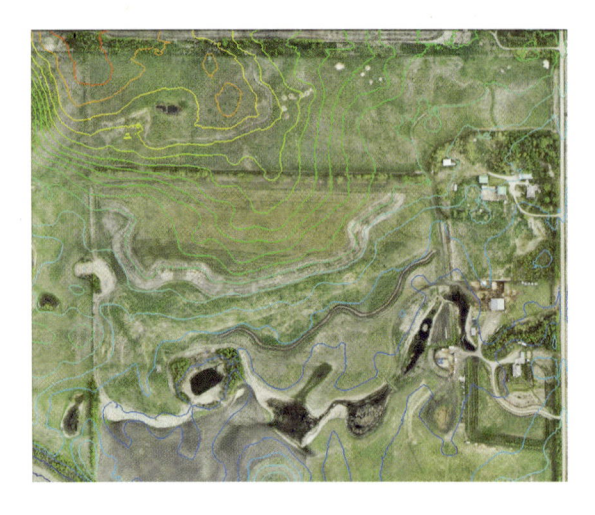

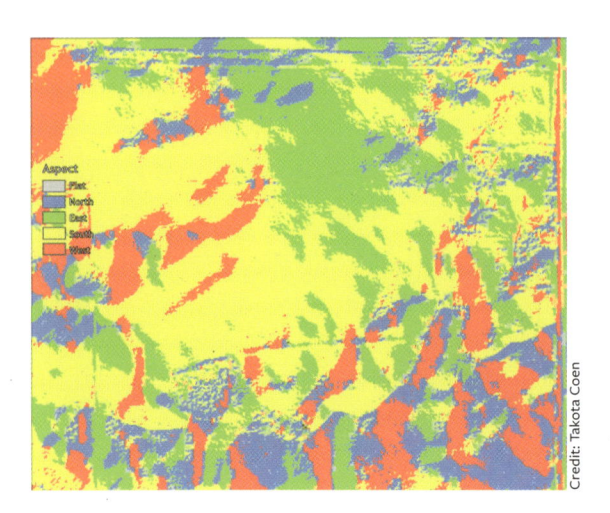

through climate, water, access, structures, fencing, flora, fauna, business, technology, and soil. Refer to Table 2.1 (or any of the templates you downloaded in practice 2.1) for more detailed subcategories and further nested subcategories.

Practice 2.3: Complete a SWOT on Your Personal and Property Resources

1. Download Template 2.3a: Personal Resources SWOT Diagnosis, create your own based on the sample provided in Table 2.4 or find a generic template online.
2. Review your resource inventory (practice 1.2) along with your vision and values one-pager (practice 1.3).
3. Complete the personal resources SWOT diagnosis by making notes about the strengths, weaknesses, opportunities, and threats you see in your existing personal resources, relative to the vision you have outlined.
4. Download Template 2.3b: Property Resources SWOT Diagnosis, create your own based on the sample provided in Table 2.3, or find a generic template online.
5. Review the information that you have organized and gathered in practice 2.1 and practice 2.2.
6. Complete the property resources SWOT diagnosis by making notes about the strengths, weaknesses, opportunities, and threats you see in your property resources, relative to the vision you have outlined.
7. Discuss your insights with your accountability partner.

FIGURE 2.9. Three of the many layers of information that you can import into GEP during the diagnosis step (contours, watershed, aspect). Learn more about accessing your own map layers at mypermacultureproperty.com.

Our Google Earth Pro Filestream

If you plan to use a GIS program to design your property, head to our website (mypermaculture property.com) to view what tools we have available for the 5SP Process for GIS. For instance, by downloading and importing the appropriate filestream into your project, you'll automatically have a detailed folder structure to build your project upon. Within your software program, the filestream not only provides the workflow (i.e., where to start and where to end), it also sets the order of the layers (and sublayers) within the stack of geographical information. Look back to Figure 2.8 to visualize this.

There are many different GIS programs, and the rate of change in the software industry is absolutely astounding. New options are constantly available to users, and current options continuously disappear. At the time of writing, my preferred program, and the one I recommend to students, is Google Earth Pro (GEP) on desktop. It's free, relatively easy to use, and provides nearly all the functionality you need as a permaculture designer. As well, there are tons of online tutorials on how to use it. As such, we've built a *5SP Process Filestream for Google Earth Pro* and make it available for download on our website. But recognizing the emergent and evolving nature of software and technology, do visit our website as we'll keep it up to date as the GIS options (and our recommendations, and subsequently our tools) change as well.

Collecting Information, Map Overlays, Open Data, and Tools for Google Earth Pro

If you are going to use GIS software, I recommend that you download a GPS app for your mobile device or smartphone. Two options at the time of publishing are GPS Kit (for Android or iPhone) or GPS Essentials (for Android). Whatever software program you choose, look for one that can export files into an output format suitable for Google Earth Pro (or whatever GIS program you are using). If need be, you can use a service like gpsvisualizer.com to convert geographic data from one format to another—but this can quickly get complicated and confusing.

Although loads of online and open-source databases contain useful and valuable information for land stewards and permaculture designers, we quickly discovered in our own teaching and consulting work that getting this information imported into GEP was a challenge if you didn't have a background in geomatics. Driven by the need to create overlays (another word for map layer) for easy import into GEP for our clients and students, my team and I built a custom web-based software service that generates multiple useful overlays based on open data. These overlays are an absolute cinch to bring into your project in GEP. This includes elevations, hillshade, contours, aspect, slope direction, river basin maps, and local catchment.

Learn more about accessing this software service on our website, in addition to suggestions on where you can acquire other open data including maps, imagery, and more.

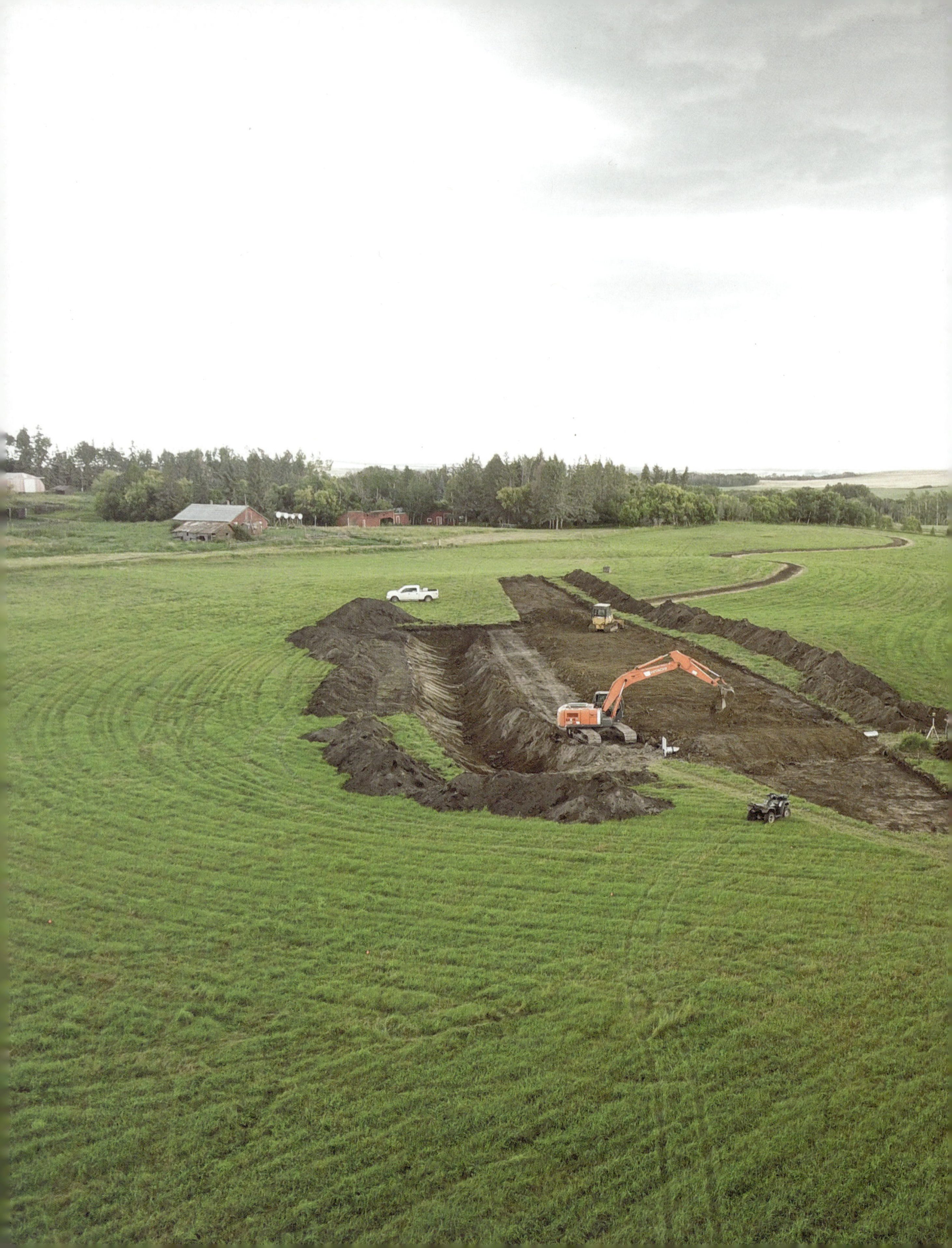

Step 3: Design Your Resources to Meet Your Vision and Values

Design

de·sign

/dəˈzīn/

verb

To devise for a specific function and/or goal.

Credit: Takota Coen

FIGURE 3.1. Once these eleven tributaries begin filling with information, they will inevitably flow downstream and deposit themselves as design ideas that start with broad brushstrokes down to minute details.

Credit: Jarett Sitter

I have heard it said that if an expert woodsman were given just five minutes to cut down a large tree or else pay with his life, he would spend the first three minutes sharpening his saw. Many other folk sayings rightly give credence to the importance of preparation in getting a job done, and building your permaculture property is no different. In fact, when Takota and I teach an online course about how to use the 5SP Process, about 70% of the time is spent on Clarify and Diagnose. The remaining 30% is more than enough to cover Design, Implement, and Manage. This always comes as a shock to our students who had expected to spend the majority of the course in design, only to have us tell them on the first day that teaching humans how to design is like teaching a bird how to fly.

As you saw in Step 2: Diagnose, humans are born with the ability to connect data points in novel ways to improve their well-being. Figure 3.1 illustrates an idea I presented earlier, that the dendritic pattern of the river represents diagnosis and the dendritic pattern of the delta represents design. If you get organized in Step 2, collect a filestream of information, and then perform your SWOT diagnoses, you will inevitably start to see opportunities and get ideas about how to best leverage your resources to achieve the well-being that you've defined in your vision and values one-pager. It is not too big of an exaggeration to say that when diagnosis is done correctly and diligently, design flows out of that work, and jumps off the page right at you. I've experienced this time, and time again. As such, this chapter simply provides you with a few additional ways to think about and organize your design.

However, while preparation is critical to your success in both cutting down trees and designing property, there is another, at least equally important factor: the quality of your tools. Put simply: preparation doesn't count for squat if your tools fall apart halfway into the job. If you are still dealing with the third struggle of permaculture—I don't know how it all fits—and you are confident that you have properly prepared in Step 1 and 2, then I guarantee you need some better tools to help you figure out how the dots connect.

By the end of this chapter, you will learn:
- What design is not, including three common misconceptions.
- The most common design mistake that I see.
- How the four variables of pattern are far more than just about diagnosis.

> **It is not too big of an exaggeration to say that when diagnosis is done correctly and diligently, design flows out of that work, and jumps off the page right at you.**

- The difference between the good, the bad, and the ugly of design.
- Why digital design tools are the best way to assemble all the information you gathered into a highly functional, flexible, and adaptable property plan.
- How to make best use of your brain.
- How to use five simple questions and an energy flow analysis to evaluate your designs and inspire further insights.

Why Design?

One of the greatest pleasures of working one-on-one with land stewards to help them build their permaculture properties is providing them with high-quality digital topographic maps and imagery. The reason I relish this part of the work is that most landowners have never seen recent imagery let alone accurate contour data for their properties. Their reactions upon seeing their maps for the first time always amaze me. When I make the big reveal, after a few moments of jaw dropping silence, I have come to expect one of two words to come out of their mouths. The first is some variation of Wow! as if they were seeing their property for the first time and hundreds of design connections flash into their minds. In fact, I still remember

FIGURE 3.2. Contour lines layered over a model of my own property, a 65-hectare (160-acre) permaculture homestead located in central Alberta, Canada.

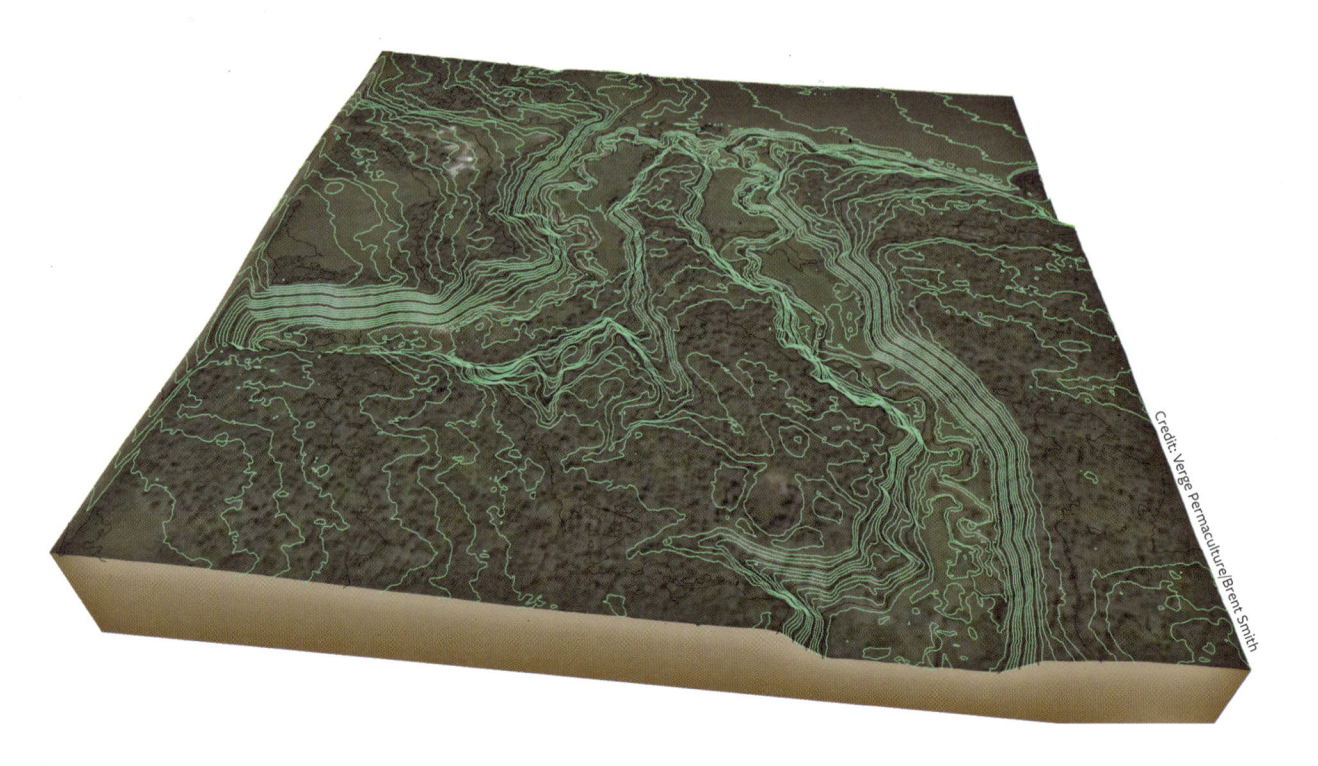

Credit: Verge Permaculture/Brent Smith

the first time I saw my own property's contour maps as clearly as my first experience of love at first sight.[2]

The other word is half as common but twice as enjoyable. Shit!—or some other four-letter word—tells me immediately that they wish they had seen these maps before they had started implementation. My personal favorite of the second type came during a consulting engagement with a goat farmer named Grant, who just months before hiring me had spent considerable money and time building a fence for his goats that was several hundred meters long, and was, as he phrased it, "bull strong and watertight." This fence was built to replace a boundary fence that had reached the end of its life and was now overgrown with thick brush. During my first site assessment of the property, before I had procured topographical maps and aerial photographs, he proudly gave me a tour of his new impenetrable wall and explained how he had to cut the existing rusted wire into short chunks where it had grown into the trees and shrubs so it could be removed by hand before starting the even more tedious work of blazing a trail wide enough for a tractor and post pounder to drive through what looked like a jungle. I commented on how much work it would have been, and he sighed happily saying, "Yup, but at least I know I'll be dead in sixty years before anyone will have to do it again!"

A week later, when I revealed his maps, the first words out of Grant's mouth were a stream of the four-letter type. Turns out, he had built his impenetrable wall in the wrong spot! For some reason, the previous property owner had built the old fence inside the actual boundary of the property. And by following the old fence line not only did Grant fence himself out of several acres of his own land but he also had to clear brush to build his fence when the neighbor's cattle had spent decades keeping the actual property line clean of anything but grass. A job that should have taken a few days ended up taking weeks. The cost of the project had more than quadrupled when taking time and cost of lost land into account.

Grant's anger didn't last long though because within a few hours diagnosing his property with his new contour maps, we were able to find almost a dozen dam and pond sites that would allow him to develop a gravity water system, flood irrigation, and livestock access network. In turn he would gain access to dozens more acres of land that was previously difficult to manage to its full potential. Later

The architect's two most important tools are: the eraser in the drafting room and the wrecking ball on the site.

Frank Lloyd Wright,[1] architect

in the design process, we were also able to turn his fence problem into an opportunity because it provided an excellent exclusion that kept Grant's goats away from the food, forage, and shelter trees he wanted to establish around his perimeter.

The story here illustrates two reasons design work is so valuable. The first is that pencil lines are cheaper than fence lines. Seeing your ideas outside your own imagination, particularly from a bird's-eye view, allows you to catch silly mistakes and fix them with nothing more than a few mouse clicks or swipes of an eraser, instead of weeks of back-breaking labor. And while the map is not the territory, it is the only way you can see the territory in its entirety. Said another way, in order to see the forest for the trees, you have to actually be able to see the whole forest. Design becomes intuitive when it follows a thorough clarification and diagnosis covered in Steps 1 and 2. Just like Grant, when you lay out all your dots, your brain cannot help but make incredible connections, and catch obvious mistakes.

And the second reason design is so valuable is, as one of my mentors used to say: "If you can draw it, you can build it." When you draw out your designs in detail, you will have valuable insights for the most efficient, effective, and ethical way to implement it in real life. You don't want to be pondering where to put your spoil pile from a pond when you have an excavator running and costing you $200 per hour. There is a saying that permaculture design is about spending one hundred hours thinking and one hour working, as opposed to the opposite. But from experience, this statement doesn't truly capture the incredible dividends of good design in its ability to minimize errors and optimize benefits. I've often wondered if the saying should be rephrased as such: "Permaculture design is about spending one hundred hours thinking and saving hundreds of dollars per hour instead of one hour of work that costs you hundreds if not thousands of dollars."

What Design Is Not

I've already argued that design is the natural and inevitable outcome of doing proper clarification and diagnosis work in Step 1 and 2. But because many, if not most, skip these two steps, much of the "design" out there results in unnecessary suffering. But apart from not doing the proper clarification and diagnosis work, there are three additional reasons why you might be struggling with how everything

> **And while the map is not the territory, it is the only way you can see the territory in its entirety.**

fits together. And they all have to do with misconceptions about what design actually is.

Misconception #1. Design Is Static

I've seen students take a permaculture design course and believe that in two weeks it is possible to create a final design for their property, then spend the rest of their life implementing it. I've also had clients approach me and ask me to deliver them a one-time master plan with this same misunderstanding. In reality, a design is not a static piece of *paper*. It is a living, breathing, iterative, adaptive, ever-changing *process*. It is a journey, not a destination. It is a verb, not a noun.

Misconception #2. Design Is Only About Aesthetics

There are many popular professions with the word "design" in their name, for instance: graphic designer, interior designer, fashion designer, landscape designer, and set designer. The primary driver of these common professions is for their work to be visually appealing. As such, this leads to the tendency to think that design, and being a designer in general, is about making things pretty. While beauty can be an end in itself that adds immensely to our spiritual resources, when it becomes the only end, the effects are as vain as they are disastrous. Or as Buckminster Fuller put it, "When I am working on a problem, I never think about beauty, but when I have finished, if the solution is not beautiful, I know it is wrong."

Misconception #3. The Best Way to Improve Is Always to Add

The third misconception is that design is always a process of addition rather than one of subtraction. Many of the permaculture designs I have seen, including some of my early work, remind me of the nursery rhyme about the old lady who swallowed the fly. And then a spider to catch the fly, then a bird to catch the spider, and so on until, before you know it, she swallowed a horse that led to her death. The snowball effect of depending upon larger and more complex elements can be lethal, or at least expensive. Think back to the story of Michelangelo and the well-known aphorism: "Perfection is finally attained not when there is no longer anything to add but when there is no longer anything to take away." Building a permaculture property is more like chiselling away at a slab of stone than adding paint to a blank canvas.

> **Building a permaculture property is more like chiselling away at a slab of stone than adding paint to a blank canvas.**

The Most Common Design Mistake

While these three misconceptions certainly contribute a great deal to bad design, the most common, and arguably the most damaging, mistake is very similar to something we discussed earlier in Step 1: Clarify. That is, the mistake of *focusing on what over why*. To illustrate this, there's an exercise that I like to do with my students, and it starts with handing out a single sheet of paper to everyone and saying, "Get this paper to fly to the other side of the room." Everyone begins folding their paper to make various airplane shapes. I then ask them to line up at the back of the room, where they launch their aircrafts with a flick of their wrists. The best go halfway across the room, but most barely make it a few steps away. At that point, I take a piece of paper, crumple it into a tight ball and toss it clear across the room, hitting the other wall with ease. The point here is that the word "fly" instantly makes us think of an airplane. Then the solution is based on the concept of an airplane (e.g., *what*). However, when we shift our focus to understanding the outcome, the function, or asking the question *why*, we can often find better or more appropriate solutions—such as a crumpled-up piece of paper!

I see this common mistake with students and clients alike, especially in our age of digital information abundance. They see, read, or hear about a specific thing that has helped a specific person, in a specific place, under specific circumstances, achieve a specific outcome. It is only too easy to become fixated on that specific "what" as the only magic pill or silver bullet that will help them achieve their desired outcome. So how do you avoid the *what over why* mistake when you are convinced that a passive solar greenhouse, swale, pig, or food forest is really what you need? The solution is to ask yourself two questions (and note how similar these two questions are to the questions posed in Mistake #1 that we previously covered in Step 1: Clarify):

1. What am I trying to achieve?
2. Are there any other options that deliver the same outcome?

The good news is that if you've followed the process so far, you've already been primed to think this way. The 5SP Process in itself is designed to help you avoid the trap of indiscriminately copying someone else's practices or prescriptions and instead develop your own tailor-made solutions.

> **The 5SP Process in itself is designed to help you avoid the trap of indiscriminately copying someone else's practices or prescriptions and instead develop your own tailor-made solutions.**

TAKOTA'S STORY: To Swale or Subsoil?

I have made many mistakes because I was focusing on the "what" instead of the "why." One of the more costly of these started with an insight while diagnosing the flora on my property. Looking back at twenty-five years' worth of crop records, it became painfully obvious that all our pasture and hayfields had slowly declined in productivity by up to 75%.

While researching design solutions for this weakness, I came across encouraging reports from credible sources about the miraculous pasture improvements gained from subsoiling—the practice of making a particular pattern of deep vertical rips into the soil with special shanks to allow rainfall, oxygen, and root penetration deeper into the soil. Being a typical farmer infected with what is known as "iron disease," I quickly rationalized the purchase of yet another piece of equipment. Eight thousand dollars and a few days later, and I was subsoiling my pastures.

It was tediously slow work. Every five minutes, a shear pin would blow when a shank hit a rock. This required getting off the tractor, hunting for the shattered shear pin in the grass, resetting the plow shank, hammering a new seven-dollar shear pin back in place, and then climbing back in the cab. I spent almost as much time out of the tractor as I did in it. But that wasn't even the worst part! Our soils have particularly sticky pockets of clay, and upon hitting this **gumbo**, the front end of our 7-metric-ton, 150-horsepower tractor would lift clear off the ground. I almost had to clean out my shorts more than once!

The original plan was to do the whole 100-hectare/250-acre farm, but realizing the high fuel, shear pin, and time costs involved, I decided to run test plots in various pastures and at various depths and widths to see if the extraordinary claims of subsoiling would hold true for our farm. A year later, the results of this experiment were in, and they were disappointing. Not one of our plots showed any increase in productivity.

It was clear that I needed to figure out "why" my pasture productivity was declining before I decided "what" I was going to do to fix it. To start this diagnosis process, I decided to start by understanding "why" the subsoiling didn't work. If the function of a subsoil plow was to create a deep vertical rip into the soil to allow water, oxygen, and roots deeper into the soil, did I have problems with rainfall, oxygen, or roots getting into our soils?

I walked around our entire farm with a soil penetrometer and determined that there was no compaction to be found, except in the few areas with gumbo. But it appeared that deep ripping here was like running a knife through jello, it just sealed itself back up. I dug test pits with a shovel and found that plant roots were already extending well beyond the depths of the subsoiler shanks. I completed rainfall simulation tests and discovered that our soils could infiltrate several inches of rain in a few minutes, a rate well above any of our extreme rainfall events recorded back more than 30 years and with accounts from my grandfather going back decades more.

After much additional diagnosis and a few other failed experiments, I came to realize two new possible reasons why my pasture production was being limited. The first was that the previous owners had mined the soil of carbon, nutrients, and minerals in the form of exporting hay, grain, and straw off the farm at rates of up to 8,400 kg/hectare/year (7,500 lbs/acre/year).

The second was that we did lack water to grow our pasture, but it wasn't a lack of rain, it was a lack of snow! Approximately 50% of our yearly precipitation comes in the form of snow that melts in less than a two-week period in the spring when our soils are 1 to 4 meters (4 to 7 feet) deep. Well beyond the depth of the 18-inch shanks of my plow.

With this new diagnosis of poor soil health and snow harvesting, I had some new design ideas about what to do. But first I had to sell the subsoil plow to free up some cash. I received dozens of calls from other wide-eyed farmers looking for the same silver bullet that I was, but I could not sell it in good conscience without sharing my experience and making sure they met the criteria of poor rain infiltration, prevalent hardpan, minimal rocks, and most importantly no gumbo! It took over a year and a $2,000 reduction in price before I found a suitable buyer.

Fast-forward five years, and I've had great success addressing the real reasons why our pasture was failing. I began by improving the soil health in our pastures with cover crops, compost, and proper rotational grazing. I also began slowing, spreading, and infiltrating the massive amounts of spring snowmelt by using swales. Unlike the small rip lines from the subsoiler, these earthworks were large enough to harvest enough water back to either pull frost out of the ground or be held back in large dams or wetlands until the frost is gone and it can be flood irrigated into the soil via swales. In addition, I have begun to develop strategies to harvest even more snow with shelterbelts and taller pasture species. With these changes, I am happy to say that our yields are increasing for the first time in decades, and in some pastures, we are even exceeding previous record yields.

Of course, swales are not perfect for every context either. Many of my clients and colleagues have had excellent results with subsoiling and terrible results with swales. If you are ever getting less than satisfactory results because you have mistaken the what for the why, there is a good chance, as my brother used to say, "It's not the tool, it's the tool using it."

Step 3: Design Your Resources to Meet Your Vision and Values 115

FIGURE 3.3. (*above*) Coen Farm swale during the spring runoff and (*below*) later that year during the worst drought in decades. Notice the significantly greener pasture on the downhill side.

Form, Timing, Placement, and Scale

Remember the wooden ball maze, Figure 2.6, from earlier? The playing field, a maze littered with holes, represents your permaculture property, and the four directions of tilt each represent one of the variables of pattern: form, timing, placement, and scale. With that in mind, Table 3.1 shows the relationship between the variables of pattern and how technique, strategy, planning, and design are defined here. When you set out to do or design something, and you only concern yourself with its form, I call this a technique. If you consider both the form and the timing, you are now employing a strategy (doing something at a specific time). When you include placement with the other two variables, you are now planning (doing something at a specific time, in a specific place). It is only when you start playing the game with all four variables that you are actually doing design. Design is doing something, at a specific time, in a specific place, in a specific amount.

The Good, the Bad, and the Ugly of Design

With having just defined design as doing something, at a specific time, in a specific place, at a specific scale, I'll now reiterate that permaculture design is therefore about trying to do the *right* thing at the *right* time in the *right* place in the *right* amount. If you achieve these things, all the while meeting the three permaculture ethics (as discussed in Step 1), then I call this a *good design*. If your design is unethical and if it causes suffering, then no matter what, when, where, or how much you do it, I call it ugly design. Lastly, there's bad design. Bad designs are ethical, or at least are trying to be ethical, but at least one of the four variables are wrong.

> *A thing is right when it tends to preserve the integrity, stability and beauty of the biotic community. It is wrong when it tends otherwise.*
>
> Aldo Leopold,[3]
> author and ecologist

Good design is what you need to aim for in your permaculture property. You can achieve this by paying attention to form, timing, placement, and scale combined with a continuous evaluation of what is ethical.

TABLE 3.1. Technique versus Design. The difference between a technique, strategy, planning, and design is based on how the designer addresses and considers each form, timing, placement, and scale.

Variables of Pattern		Technique is…	Strategy is…	Planning is…	Design is…
	Form	Doing something.	Doing something	Doing something	Doing something
	Timing		at a specific time.	at a specific time	at a specific time
	Placement			in a specific place.	in a specific place
	Scale				in a specific amount.

Good design is what you need to aim for in your permaculture property. You can achieve this by paying attention to form, timing, placement, and scale combined with a continuous evaluation of what is ethical. Let's further explore the good, the bad, and the ugly of design with a few examples from Coen Farm.

TAKOTA'S STORY: The Good, the Bad, and the Ugly

There's no better way to clearly point out the difference between ugly versus good design than by juxtaposing two very different livestock management systems. I can think of no better example of ugly design than that of the Confined Animal Feeding Operations (CAFOs for short). Let me provide a sketch of what this ugly design entails in direct comparison to good design of the integrated livestock system on Coen Farm, which I also call my "hub and spoke system," illustrated in Figure 3.4. As you are reading, see also if you can identify differences in form, timing, placement, and scale, noting that some statements touch on several variables of patterns concurrently.

CAFO: Takes an enormous number of a single species of animal like chickens, pigs, or cattle, and locks them in cramped conditions inside a climate-controlled building that requires constant use of energy for ventilation, heating, or cooling in order to keep the animals from suffocating, freezing, or cooking themselves to death with their own body heat.

Coen Farm: Integrates an optimized number of chickens, pigs, and cows in an outdoor free-range polycultural system. A minimal amount of buildings provide multiple functions made possible by smart envelope designs for passive cooling, ventilation, and heating. For example, the chicken house uses the body heat of the chickens to not only keep the building above freezing in the winter but also ferment the feed for the pigs and chickens year round.

CAFO: Pumps the chickens, pigs, and cows full of vaccinations, anti-parasitics, and other pharmaceuticals to keep sickness and mortality to a minimum.

Coen Farm: Uses proper nutrition, rotational grazing, and the chickens, pigs, and cows themselves to help break each other's pest and disease cycles. Hasn't used vaccinations, anti-parasitics, antibiotics, or any other drugs for 30 years. Uses natural medicines like garlic, wormwood, essential oils, and diatomaceous earth if needed.

CAFO: Focuses on bringing the food to the animal. They feed the chickens, pigs, and cows a ration of soy, corn, grains, and sometimes animals and manure from their own species that are low in nutrient density and diversity but high in protein, carbohydrates, fats, and sometimes even growth hormones formulated to encourage rapid growth rates.

FIGURE 3.4. The integrated livestock system on Coen Farm, also the "hub and spoke" system.

Coen Farm: Focuses on bringing the animal to the food. The chickens, pigs, and cows are encouraged to freely forage for their own food in a variety of ecosystems year-round. This allows the animals to eat each other's manure and butcher scraps if they choose, but never of the same species. Any supplemental feed is of the highest nutrient density and diversity, free from any growth hormones. Harvesting of the animals is done on-farm in a zero-stress environment. Animals grow at a natural rate, and breeding animals die a natural death (often living 3 or 4 times longer than their CAFO counterparts).

CAFO: Clears forest or plows up native prairie to grow monoculture crops for the animals on other farms great distances or perhaps even oceans away. Uses large amounts of fossil fertilizers, biocides, and genetic engineering.

Coen Farm: Uses perennial polycultures like pastures and forests as the primary food source for the animals. Uses zero synthetic fertilizers, biocides, or genetic engineering. Supplemental feed is grown organically primarily on-farm or on other organic farms located as close as possible to minimize embodied energy in the grain. There is a conscious effort to eventually eliminate the use of grain altogether through the development of better genetics and alternative feed sources.

CAFO: The combination of high concentrations of animals, inappropriate diet, and toxic residues from pharmaceuticals and biocides leads to the buildup of animal manures that pollute aquifers, streams, lakes, and oceans.

Coen Farm: A properly tuned carrying capacity of all animals being fed appropriate diets free from pharmaceuticals and biocides improves the health of the aquatic and terrestrial ecosystems. The problem of nutrient runoff is transformed into a solution in the form of a duckweed and minnow harvesting system that produces high-quality protein. The duckweed and minnows grow rapidly on the livestock manure and are safely fed back to the chickens and pigs whose manure was used to create it (more on the duckweed and minnows later).

It is clear that industrial animal agriculture does too much of the wrong thing, for too long, in the wrong place. It is a completely wrong design that leads to nothing but suffering, pollution, and drudgery. While CAFOs spur an appropriate response from ecological and animal rights activists, they also misguide and lead to blanket statements from some vegetarians and vegans. It's not necessarily the chicken, the turkey, the pig, or the cow; it's the what (form), the when (timing), the where (placement), and the how (scale)!

Bad Design

So what is stopping a bad design from being a good one? As long as it's ethical, the difference is simply a miscalculation in one or more of the four variables of pattern. And more often than not, even on my own farm, designs start out as bad, and then require iteration and tweaking before they are good! Here are a few personal examples.

Wrong Form

Excited about the incredible productivity of aquaculture systems, I decided to experiment by adding some rainbow trout to a new dugout. Taking into account recommendations from my local agricultural department, I made the dugout 6 meters (18 feet) deep to make sure there was adequate water to not freeze solid during our frequent 40° below zero cold spells. I made the side slopes as steep as possible to prevent wading predatory birds like blue herons from making an easy lunch out of my fish. I also purchased a decoy blue heron, believing that it would deter other herons from encroaching on its territory. And just to be extra safe, I acquired an inflatable water snake and alligator! A windmill aeration system was added to improve the water quality and oxygen levels. To eliminate the need for soy-based fish food, I even added an insect-attracting raft in the centre of the dugout complete with solar-powered lights shining on the water and fragrant colorful plants

FIGURE 3.5. (1) Morning chores in the hub of the integrated livestock system. (2) Takota and his animals out in the forest garden spoke of the integrated livestock system.

FIGURE 3.6. (1) Rock Barred chickens eating three-month-old fathead minnows. (2) Rainbow trout after twelve months of growth.

suspended from the sides of the raft to maintain a consistent depth in the water regardless of the water level in the dugout.

After purchasing an aquaculture license, I went to my local fishery and purchased one hundred 15 cm (6-inch) rainbow trout fingerlings and introduced them to their new home. All of this added about $1,000 in extra costs to the dugout, but if these trout grew to the reported 5 kilos (10 pounds) each within a few years, I would be able to recoup my costs, and make a handsome profit, by selling them for $20/kilo ($10/pound), the same price I sell my grass-fed beef for.

Right off the bat, things started to go wrong. While introducing the trout, I noticed many of them were infected with a parasite known as whirling disease. Within days of stocking, a giant blue heron never before seen on our farm had promptly knocked over my plastic decoy and claimed the territory for himself. Finding the side slopes too steep for wading, he was using the raft as a landing pad. And because of the way I had designed the raft to attract insects, it was ideal feeding grounds. My snake and alligator were nowhere to be seen, but I wouldn't be surprised if he ate them too!

After a stressful summer of unsuccessful attempts to scare away such a beautiful bird that I would have normally welcomed, I found some solace in the odd splash that indicated I still had some fish left. Later that winter, fearing they would all be eaten before they reached their target size, I set to work in an elaborate attempt to set a gill net under the surface of the ice. A three-day affair of chainsaws, ice jiggers, and frozen hands yielded nine fish, all under half a kilo (one pound). They were good, but they weren't that good.

The last straw was the following spring when the remaining trout washed up on the shore after the ice melted. Low wind conditions meant that the windmill didn't do its aeration job, and my fish suffocated during oxygen depletion from pond turnover.

Looking back now, it is obvious to me that these fish would not have survived and thrived in a static lake-style environment with lower water quality than their natural highly oxygenated river ecosystems. Not to mention the fact they were sourced from an industrial farming operation that purposefully sterilized all the fish so they could never reproduce. But I took these risks because trout is the only fish I was legally able to stock without facing a $100,000 fine and or a year in prison.[4]

Luckily though, when you create a reasonably good design, nature takes hold and makes it better. A year after all my trout had died, and I hadn't bothered to try again, I started to see fish jumping in the water again! Native fathead minnows (*Pimephales promelas*) had somehow swum upstream during the spring melt that now, as a result of extensive water harvesting efforts, connected us to our greater watershed where they were abundant. These incredible little fish can lay more than 10,000 eggs per year and reach their full size of 7 centimeters (3 inches) in 90 days; they thrive in poor-quality low-oxygen water and are very easy to trap at scale.

My pigs, chickens, dogs, cats, and even some of my human customers love the taste of these little fish, and they are now an integral part of the developing aquaculture systems on our farm. Best of all is the fact that I am now rightfully joyful when I see yet another blue heron circling our farm looking for a place to land.

If you are having trouble growing a particular plant or animal, consider changing its form. I have had excellent results trialling different varieties of the same species of flora and different breeds of the same species of fauna. If that doesn't work, consider finding other analogue species by shifting the genus but keeping to the same family. You will find that certain forms of organisms will flourish with zero help and others fail utterly.

Wrong Timing

A few days after teaching a daylong summer workshop about my integrated livestock system, I received a call from one of the participants, Sarah, who asked me rather timidly, "Have you ever had your pigs eat your chickens?" Sarah was so excited by the success of my system she had promptly integrated her pigs and chickens the day following the workshop. Her pigs, in turn, promptly ate several dozen of her chickens. I was shocked. I had never seen the slightest sign of predation from my pigs, even though I regularly fed them older culled chickens—something they absolutely relished! I started my diagnosis by asking about scale, "How big was the pen, was there room for the chickens to escape?" Her pen was the same size. Next I asked about placement, "Were they locked into the pen together where the chickens couldn't escape?" Her answer was no and that her chickens could free range. Then I asked about form, "What breed of pigs and chickens?"—both were the same as mine. Lastly, I asked about timing, "What age were the pigs and chickens?" There was the answer. It turns out that Sarah's pigs were six months old (the equivalent to rambunctious teens in pig years) when they were introduced to their first flock of chickens, who were only weeks old themselves, practically babies. Having seen how my teenager pigs will prowl around like a gang of thugs vigorously inspecting any new element placed in their pen, typically by scratching themselves on it or biting it, the report of their delight in discovering soft, warm, delicious, and animate toys in their midst now came as no surprise to me. Indeed, I've even been nibbled a little too eagerly myself by pigs at this age.

Now when I teach this workshop, I make sure to explain a few of the crucial timing elements for setting up this integrated livestock system. First off, the ages of the animals is important. The first time my pigs were integrated with chickens was when they were already two-year-old sows, practically grannies who could care

Step 3: Design Your Resources to Meet Your Vision and Values   123

FIGURE 3.7. (1) One of two daily 45 kg (100 lb) harvests of duckweed. (2) Aerial view of the hub and spoke system and the duckweed pond.

less about anything other than food and sleep. From that point on, any new piglet is raised from birth with full-sized chickens who would actually attack the baby pig if provoked. By the time the pigs are big enough to do any damage, the chickens are old news, not worth investigating, and the chickens have learnt to watch their backs. Secondly, the seasonal timing may also help ensure success. My initial successful integration happened in the middle of winter, when both species were not as active.

Sarah's design was identical to mine except for timing. The good news is that Sarah tried again with these small timing tweaks and reported back the following year with a successful, and masacre-free, integration.

Wrong Placement

Duckweed (*Lemna minor*) is an aquatic plant about the size of a lentil that floats on the surface of most water bodies worldwide. After a chance discovery that my chickens loved eating it, and that it was the perfect solution to animal manure being concentrated in a centralized location because of my hub and spoke system, I started to use duckweed as a high-protein feed in my integrated livestock system for both my pigs and chickens. During prime growing conditions, I am able to harvest up to 90 kg (200 lb) per day for both my pigs and chickens, which at times makes up 50% of their diet.

When I first decided to start harvesting duckweed, I built a dock on the west side, thinking that it would be ideal to harvest on the same side of the pond as the livestock. However, within short order, I realized that the prevailing west-to-east winds meant that the duckweed was often blown against the opposite eastern side of the pond.

To get around this, I devised a casting net that could be thrown a fair distance, but it was less than ideal. After a week of this, I knew that I needed to move my dock, but I had a fair amount of resistance to the idea because it meant relocating the whole dock and cutting a gate through a fence on the other side of the pond. But most of all, I didn't want to admit to myself that I had missed something so obvious as taking into account wind patterns. So I decided to add instead of subtract. For two years, I spent hundreds of hours trying to make my design better instead of doing it differently. I built a water fountain system designed to blow water, and the duckweed, back towards me, but this took too much time and too much water. I dropped a log into the pond and tethered it halfway across the pond to act as a barrier to stop the duckweed from blowing out of reach, but this served only as a snag for my casting net. I even tried using a canoe to extend my reach, but after a few close calls almost capsizing, I gave that up.

Finally, I admitted I was wrong, and I built a second dock on the east side of the pond. All the other variables of scale, time, and form stayed exactly the same, and instantly my design was better. I was able to double the amount of duckweed I was harvesting per day, while decreasing the amount of time it took to harvest it by half!

Wrong Scale

After some lucky diagnosis work revealed a massive Black Swan threat that a nearby landfill posed to our farm's groundwater aquifer (see the story Growing Up a Carpenter in Step 2: Diagnose), I started to design a new water source to mitigate the threat of toxic leachate polluting my aquifer and the weakness of having only one reliable water source, a deep groundwater well. After a substantial number of hours of thinking, the best design that emerged was to use a solar pump to lift water to a dam located high in the landscape. I would then be able to use gravity to passively distribute water anywhere on my land. The dam I was planning to build would hold 230,000 liters (60,000 gallons) of water and would be connected to a 400-meter-long (1,200 foot) swale positioned just below the dam and running across the landscape on contour. This swale served three purposes: to divert and spread out any water overflow from the dam, to harvest snowmelt and fill the dam in the event it is empty in the spring, and to flood irrigate a food forest planted along the swale.

To facilitate these overflow, filling, and flood irrigation functions, my design of the dam wall included a 150 mm (6 inch) diameter swivel pipe through its center as shown in Figure 3.9. This swivel pipe is what allows the multifunctionality previously described. By swiveling the upright pipe section, the swale can be flooded from the dam and/or the dam can be filled from the swale, depending on their respective water levels.

The first time I excitedly went to fill the swale from my newly constructed dam, it was thrilling to watch the 30 liters per second (500 gallons per minute) of water gushing from the downward pointing swivel pipe into the swale, using no power source except gravity. Seeing how well everything was working and knowing that it would take about two hours for the dam to empty, I left to get some other work done. Upon returning, I found the dam empty, and I began wading triumphantly along the swale, with my rubber boots nearly filling with water. I was very pleased! My satisfaction, however, vanished upon rounding a corner to see that the spread of water stopped at about the halfway point along the full length. My stomach sank. I must have got my elevations wrong during construction, I thought.

But upon checking and rechecking, it was plain that the swale was perfectly on contour. It was the scale of the swivel pipe that was off! The swale was infiltrating faster than the flow rate of water from the 150 mm (6 in.) pipe because I had underestimated the ability of my soils to infiltrate water. What is more, after several years

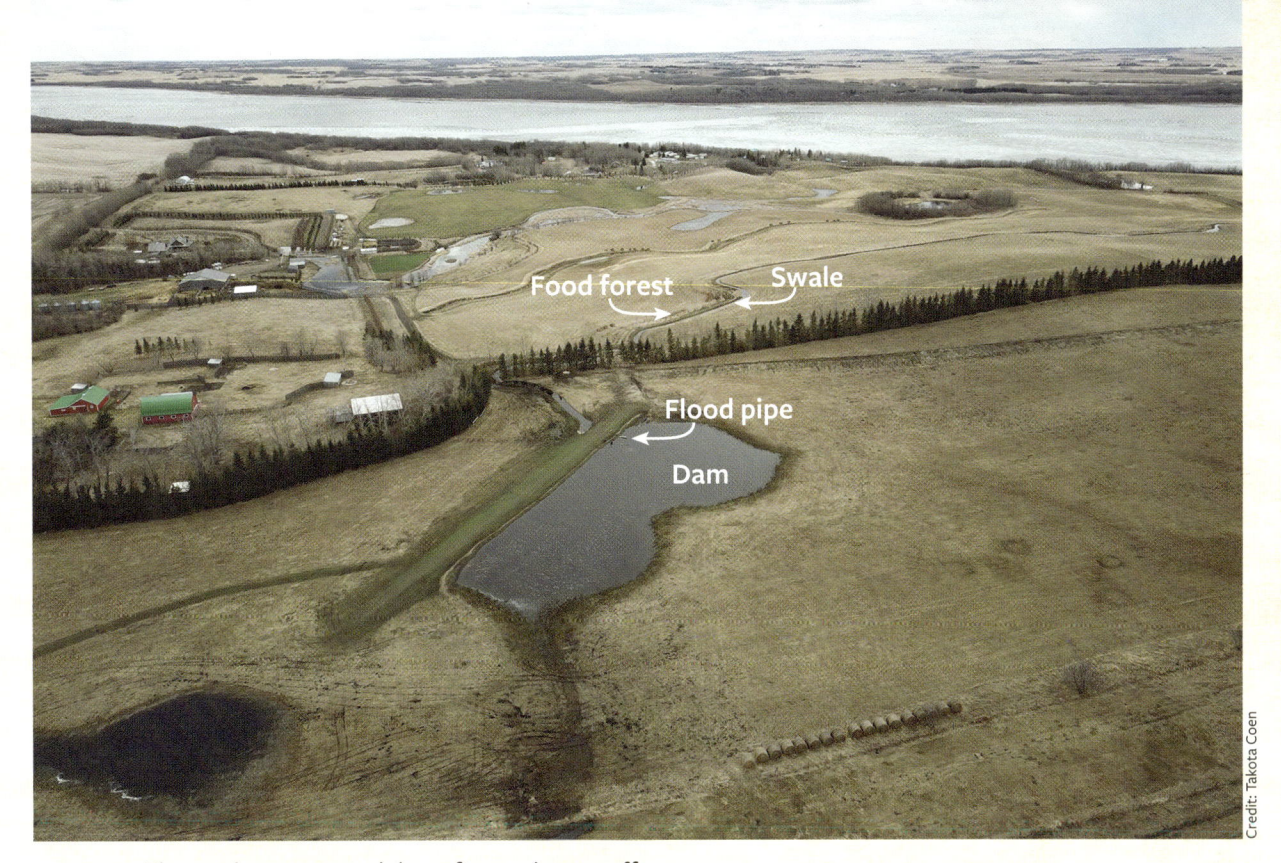

Step 3: Design Your Resources to Meet Your Vision and Values   125

FIGURE 3.8. The newly constructed dam after spring runoff.

of managing my land regeneratively, I found that the flooded length further diminished as the soil carbon further increased (more on soil carbon and water holding capacity later).

The original design had the right form, the right timing, and the right placement. But the scale of the swivel pipe and the scale of the stored water volume in the dam were wrong. Because all of the functions of this dam and swale combo worked so well, with the exception of the flood irrigation, it was a number of years before I rebuilt the dam during one of the earthworks workshops I regularly host with Rob. But this time, I used a 400 mm (16 in) pipe that gushes 250 liters per sec (4,000+ gallons per minute)—nearly a tenfold increase. In addition, I increased the size of the dam so that the total stored water volume is now 2.2 million liters (600,000 gallons)—also ten times the previous amount. With this new design, this swale not only now distributes water across its full length, it also can capture enough snowmelt to provide 100% of our domestic, livestock, and crop irrigation needs with no pumps of any kind to fill or draw water from the dam. I can at least sleep a little easier knowing our property will not be affected if the nearby landfill is leaching toxic waste into our groundwater.

FIGURE 3.9. (*top*) The 150 mm (6 in.) swivel pipe going through the original dam wall. (*bottom*) The 400 mm (16 in.) flood pipe going through the newly constructed larger dam.

There are two very important takeaways from these four examples shared by Takota. The first takeaway is that a small change in just one variable can have a significant impact on the design and more importantly on the outcome. The second takeaway is that right designs are almost always an evolution and iteration upon not-quite-right designs. Before a design is good, it usually starts out as not quite right or as bad design. This really highlights the pitfall of Misconception #1, the belief that design is static and can be created one-time on a piece of paper, and then implemented.

In the next few sections, I'll share a few more of the ideas that I have found most useful when it comes to creating and thinking about a good design.

Creating a Permaculture Design

Design is the *subject* of permaculture, and as such, there are an incredible number of fantastic resources[5] out there for the aspiring designer. Here I simply offer a few additional ideas and principles that my co-authors and I have found instrumental in our own design, teaching, and consulting work.

Make It Adaptable and Interactive

I've already recommended that you consider using digital mapping software to collect and organize your own data as well as download and import third-party information for your property (such as government historic weather data, satellite-based topographical maps, and much, much more). However, there's yet another reason why I strongly encourage you to go digital, and that is the ability to create an adaptable and interactive design. Think back to Michelangelo and his three-dimensional wax model. As discussed above, your permaculture property design is not at all a static, one-time thing. As such, although it can be done on paper and with pencil, there's no question that digital mapping tools really stand out here both for ongoing functionality and ease of updating as things change and evolve over time.

If you started your digital map and set up your filestream in Step 2, you've already built many layers of geographic information. Often, these diagnostic contour lines, placemarks, paths, polygons, photos, and text are leveraged and reused in Step 3 as you start to create an intricate, functional, and layered design plan. Layers can

> However, there's yet another reason why I strongly encourage you to go digital, and that is the ability to create an adaptable and interactive design.

easily be toggled on or off if useful to what you are considering. For instance, you might turn on the contour overlay when looking for opportunities to passively capture water in dams or dugouts. It can then be turned off when you are evaluating your fencing layout. You can also create separate overlays for existing fence lines and future/planned fence lines, all the while continuing to file both your design and planning ideas along with other information in the appropriate layers in your filestream. See Figure 3.10 for an example of how this looks in Google Earth Pro.

Set Yourself Up for Deep Design Work

Physically speaking, humans are a rather unimpressive species. We can't compete in strength with most other animals our size; our senses of sight, hearing, and smell are lackluster compared to most species, and we may be the only species that develops lower back pain from our normal[7] range of activities and posture. How then have we come to be the greatest geologic force of our time? Quite simply because of one particular region of the brain: the prefrontal cortex. From an evolutionary perspective, this rather recent addition to the brain is located directly under our forehead and is responsible for our problem-solving aptness and therefore our capacity to design. The prefrontal cortex is where we cross-reference to other parts of our brain to imagine, invent, derive conclusions, and so much more. And although design happens naturally and intuitively, it is energy-intensive work that requires you to focus intently on one problem. It turns out that there is a bit of a Goldilocks zone that facilitates our brain going into what is known as deep work,[8] where your cognitive and design capabilities are at their best.

Unfortunately, getting into a deep work state is not always easy. First, *any* distraction inhibits your capability to completely focus. Secondly, your state of mind and environmental conditions aren't always conducive to focussing intently on one problem. For instance, deep work is incredibly difficult to perform if you are anxious, depressed, or upset, or if your brain is rambling and otherwise preoccupied with an internal story.

Luckily, there are methods you can practise to find this Goldilocks zone, and perform deep work with regards to designing your permaculture property (or anything else in your life for that matter).⁹ The first task to do is schedule your time for deep work. Scheduling it will not only make it a habit but will also further save your willpower (a finite resource most abundant in the morning). Consider scheduling deep work for first thing on Monday morning, a practise Michelle and I engage in and call "Prefrontal Monday."¹⁰ Mondays are perfect for this because it is when your brain is well rested and disconnected from work after the weekend. Deep work should be scheduled for at the very least an hour, and be the only thing you work on, look at, or engage with. If an hour is too long of a stretch, consider using the Pomodoro technique: allotting twenty-five-minute slots of deep work, taking a five-minute break, and then another round of deep work. After four intervals of deep work, take a longer break. You may find that once you are in the zone of deep work, you want to stay in it and will work very efficiently for a longer than expected period.

Human beings, it seems, are at their best when immersed deeply in something challenging.

Cal Newport[6]

Variations of Two Classic Permaculture Design Tools

Out of the full permaculture toolbox of principles, tools, techniques, and ideas, there are two methods of design that I lean on for eighty percent of my design work. These are slight variations on the classic *needs and yields analysis* and the well-known **sector analysis**; if you haven't heard of either, you need do some reading on permaculture design! Here is a brief overview of how I tend to think through these two methodologies.

An Expanded Needs and Yields Analysis: Ask Five Questions

The classic needs and yields analysis is a fantastic brainstorming activity to ensure that the products of one element (the yields) are matched with the needs of another, as this is one of the goals of a self-regulating system. The way that I apply this in practice when I'm thinking about solving a design problem is to ask myself the following five questions:

1. Is every need of every element met within the system?
2. Is every yield of every element used within the system?
3. Is every element serving multiple functions within the system?
4. Is every function required by the system served by multiple elements within the system?
5. Is the system and its elements functioning ethically?

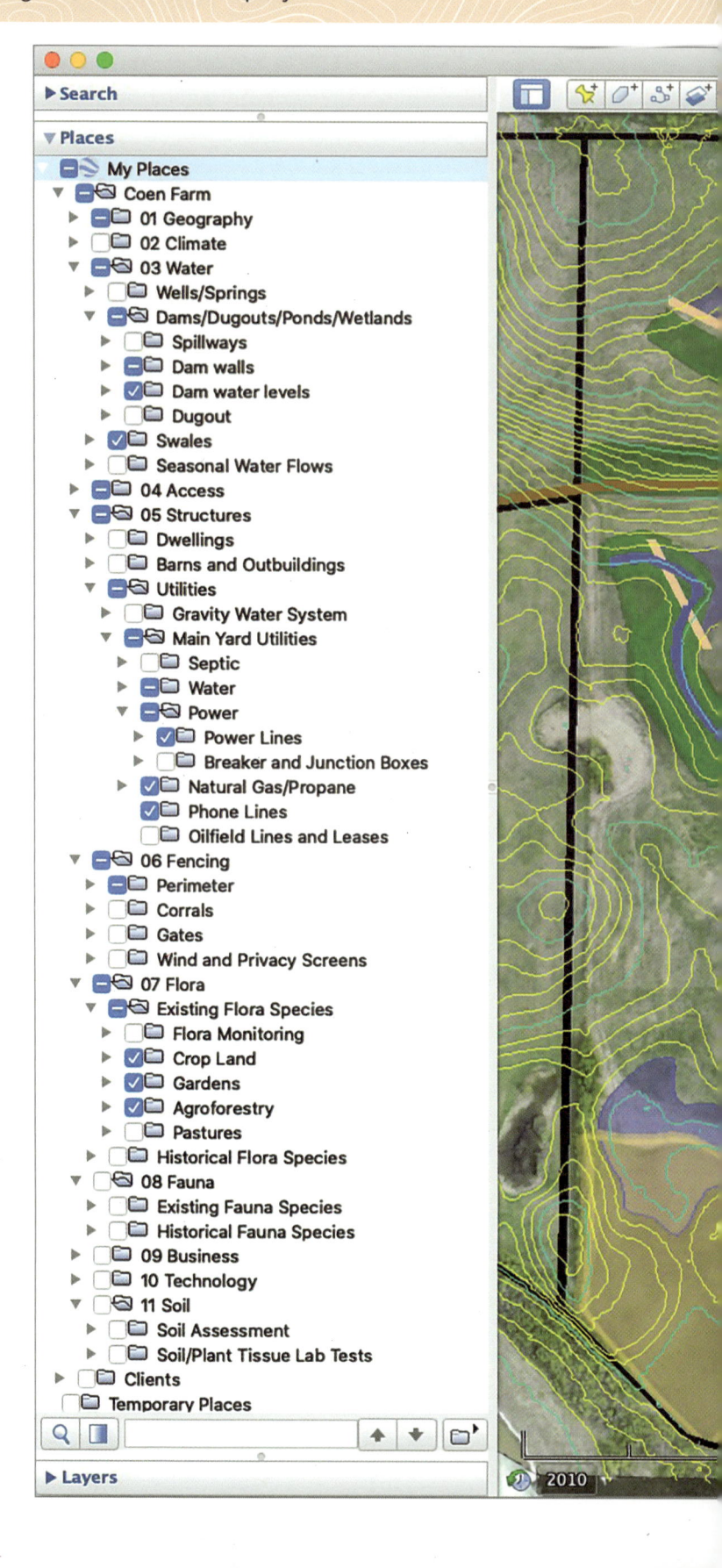

FIGURE 3.10. Coen Farm design using Google Earth Pro. Note the filestream in the Places panel, showing which subfolders/sublayers are toggled on.

Step 3: Design Your Resources to Meet Your Vision and Values   131

FIGURE 3.11. Use cycles, storages, and shields to maximize the collection of the energies you want and minimize the collection of energies you don't as these flow from source to sink on your property.

Credit: Jarett Sitter

Note that the first two questions are about analyzing needs and yields, and the second two prompt me to consider multifunctionality and redundancy. The last question is about keeping *What is right* in mind because, as you will see later in Step 5, if there is needless suffering in your system, it is a clear sign that one or more of the needs and yields in your system is unmet. I find that these five questions are a simple way to evaluate my designs, and I also use them to monitor my implementation and/or clarify what I need in order to start an upwards spiral.

A Variation of Sector Analysis: Design for Energy Flow

As far as I can tell, the universe is nothing more, or less, than a seemingly infinite amount of various forms of energies interacting with each other as they flow from their respective sources to their respective sinks. Luckily for us, there are laws that govern how these energies flow and interact. Patterns are the physical manifestation of those laws: hot air rises, water flows down hill. Snowdrifts form on the leeward side of trees, and the sun follows a set path through the sky. One of your goals as a property designer is to figure out how best to interact with these energies, and the traditional permaculture sector analysis exercise is a very useful way to do so. Sector analysis involves creating a map and visualizing patterns and energies as they "flow" across your site. I like to significantly expand on this exercise by thinking of permaculture design through an analogy of an old-fashioned pinball arcade game, as shown in Figure 3.11. The pinballs being dropped into the top of arcade game represent the **sources** (external energies or inputs), and the very bottom of the arcade field, where the ball falls out of play, is the **sink** (think of a drain, any export or energy leaving your site).

Let's consider for a moment that the design of most modern-day houses, properties, communities, cities, and so on require a continuous source and supply of external inputs to keep these systems operational. You can then visualize those external inputs, or energies, as a continuous supply of pinballs that are being dropped into the game and simply allowed to go immediately to a drain, also called a sink. Think of the rain that hits a city roof and is directed immediately to the storm sewer. Or think of the photons hitting a building or the landscape that are not used for heat, electricity, or photosynthesis.

Engage and partner with your natural surroundings and with your community-plant the rain, dance with the sun, grow the shade, feed the soil, sail with the wind, and be one of your community's spark.

Brad Lancaster,[11] author and rainwater harvesting expert

Or of food purchased at the grocery store, and "waste" that is sent to landfill. In each of these instances, the pinballs are dropped into the top and allowed to fall directly into the drain without even attempting to flip them back up using the paddles.

Bill Mollison has a famous quote: "When the needs of a system are not met from within, we pay the price in energy and pollution." As such, permaculture designers are fascinated and driven by the idea of creating **cycles** that keep these pinballs in play as long as possible. In our analogy, this is equivalent to adding paddles to the arcade field. As a real-life example, it might involve ensuring that rainwater is captured and utilized, or that all organic matter is composted and used onsite, instead of being sent to landfill.

As these examples suggest, creating cycles often requires building or expanding **storages**, which allow you to increase the **stock** or amount of certain energies you have stored on your property for future use. Sometimes, however, you might not want energies interacting with other energies in your system at all. In these instances, you could consider creating **shields** that prevent certain pinballs from entering the arcade game in the first place. An example of this might be building a windbreak or a fence to reduce strong wind or keep out wildlife.

In summary, when you design for energy flow, start by considering the various energy sources and sinks that you have on your property. Next, decide whether or not you want the energy. Then develop a design about how best to use your resources to create appropriate cycles, shields, and storages that maximizes the collection of the energies you want and minimizes the collection of energies you don't want. Table 3.2 provides tangible examples for sources, sinks, cycles, shields, and storages on your permaculture property; note that we will go into more detail, particularly on the topic of cycles, in Step 5: Monitor.

> **Permaculture designers are fascinated and driven by the idea of creating *cycles* that keep these pinballs in play as long as possible.**

Practices for Step 3: Design

Here are some practices (i.e., tools and templates) that I have found particularly useful based on the discussion in the preceding chapter. As always, you can find any of the templates or resources discussed below at mypermacultureproperty.com.

TABLE 3.2. Designing for Energy Flow. Examples of sources, sinks, cycles, shields, and storages that you may find, or design into your permaculture property.

Sources	Solar energy (heat and light), water inflows, rain, snow, wind, wildlife, wildfire, humans, petroleum products, electricity, human traffic.
Sinks	Water drainage leaving the property, culverts, property boundaries, the ground, the lowest elevation on your property, any export of biomass products for sale or trade, anything sent to landfill or offsite.
Cycles	Nutrient cycling and water cycling including: composting, soil building, graywater systems, rainwater harvesting, pumping water from low elevation to high elevation. Ecological succession: increasing biodiversity, fauna and flora increasing in size with longer life cycles. Energy cycling through waste heat recovery systems, passive heating and cooling, and financial cycling through vertical business integration.
Shields	Windbreaks, firebreaks, drainage systems, eavestroughs, specific landforms, structures, shade structures, clothing, fencing.
Storages	Biomass including flora, fauna, soil carbon, soil life. Water storages such as tanks, ponds, and soil. Energy storages such as compressed air, biofuels, wood, battery storage, thermal mass. Food storages, both living resources and preserved goods.

Practice 3.1: Set Yourself Up for Deep Work

Getting into a state of deep work is unfortunately easier said than done. There are, however, some tactics that can ease this process. The first is to schedule it in, particularly when you feel rested and happy, and stick to it.

Meditation is one of the most simple and effective practices you can do to improve your ability to perform deep work. If you have practised meditation before, you will be familiar with the mind-sharpening and increase in self-awareness that follows; if you are new to the concept of meditation, it is strongly recommended that you begin—it is nothing short of life changing.[12]

Further Tips

- Don't half-ass anything you do. When you are working, work; when you are chilling, chill. Make it a binary.
- Keep your work area tidy. A lot of clutter means a lot of visual stimulus for your limbic system to constantly scan through.
- Set up a ritual for getting into deep work. For instance, going for a 10-minute walk through the woods or a field before getting to

work acts as a segue and cue for deep work, rather than turning on a dime into it (which expends a great deal of willpower).
- Quit social media.
- Embrace boredom and use this time to further your development.
- Drain the shallows. Reduce the amount of time you are spending on shallow work tasks such as meetings or tasks that don't require all of your attention.
- Employ the pomodoro method discussed above.
- Have your accountability partner hold you accountable.

Practice 3.2: Study Case Studies

If you find yourself stuck, or need inspiration or ideas, reviewing other people's designs is a great way to gain insights and ideas. However, when you do this, there are two things to watch out for. Remember that the most common design mistake is focusing on the what over why. For instance, instead of "She used a swale, and it worked therefore I must use a swale," ask yourself "Why did she use a swale?" and, "What did this accomplish?" Secondly, remember that design is the combination of form, timing, placement, and scale with a continuous evaluation of what is ethical, efficient, and effective. If you purely copy someone else's technique (i.e., form), you are sure to get it wrong.

So, get out there, go on farm tours, attend workshops at permaculture sites, participate in work bees. Do your research by seeking and reviewing case studies both in books and through online channels. We will also endeavour to make available an inventory of case studies on our website.

Practice 3.3: Create Your Adaptable and Interactive Permaculture Property Design

Pull out your SWOT (practice 2.3), your filestream (practice 2.1), your one-pager (practice 1.3), and your resource inventory (practice 1.2). Think back to Figure 3.1 at the beginning of this chapter and remember that if you do a good job with your diagnosis, design will flow naturally. Starting at the top of the filestream with geography (see Table 2.3), work through each of the categories and subcategories and look for ways to leverage your strengths and opportunities and mitigate or eliminate weaknesses and threats. Make edits to your design, your vision and values one-pager, and your resource inventory as new insights begin to emerge.

Remember to organize your design ideas within whatever filestream format you choose while aiming to create a right design using the form, timing, placement, and scale variables. Take a commonsense approach when it comes to your budget, but don't necessarily let your current resources limit your design. Dream big, noting that we tend to hit the things we aim for. We'll get into prioritizing, implementing, and making specific decisions about your plan in Step 4.

> **Practice 3.4: Consider the Five Questions and Design for Energy Flows**

Use the five simple testing questions and the energy flow analysis discussed earlier (Table 3.2) to evaluate your designs and/or inspire further insights. Alternatively, download the following templates to use as checklists during your design process:

- ▶ Template 3.4a: Five Questions
- ▶ Template 3.4b: Design for Energy Flows

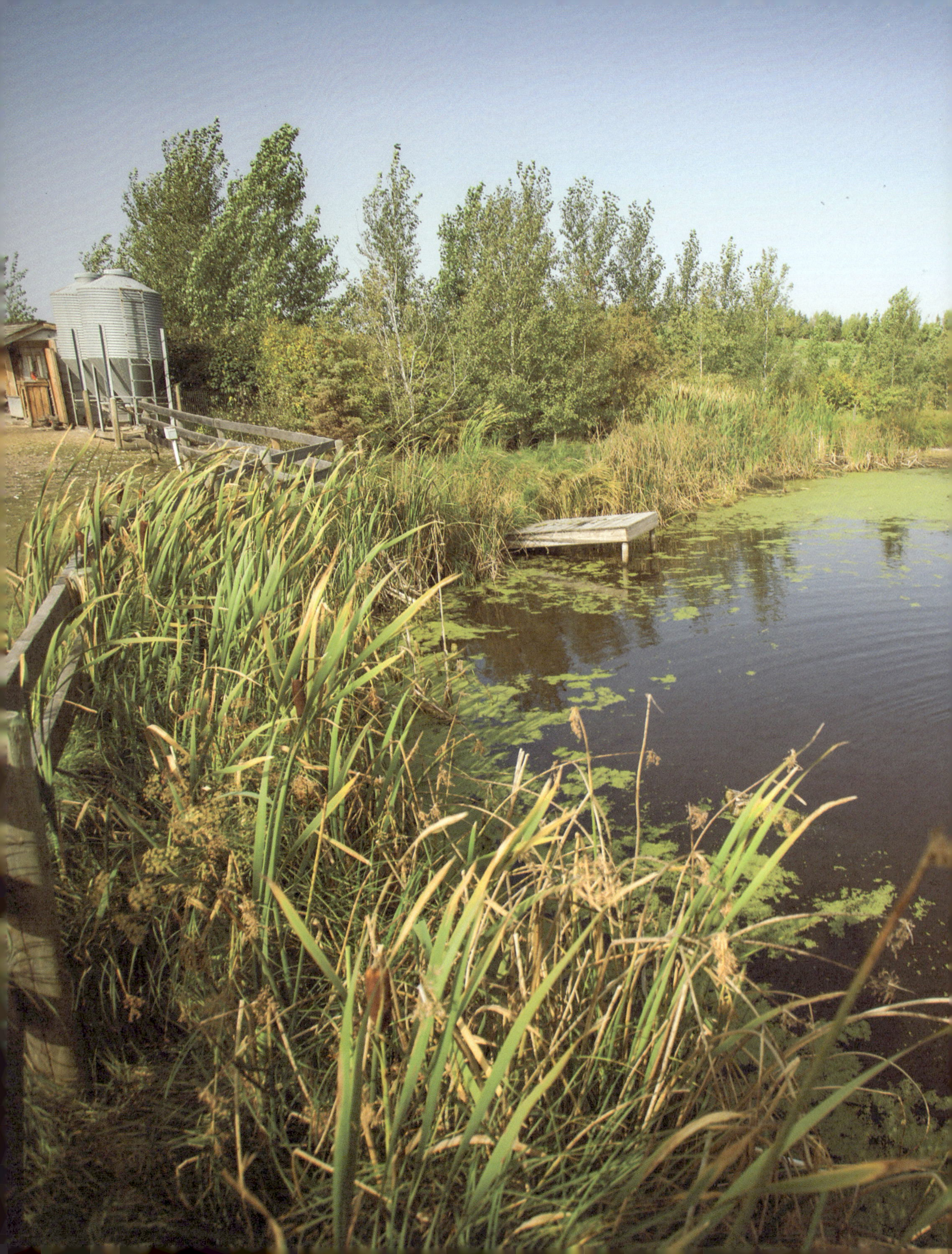

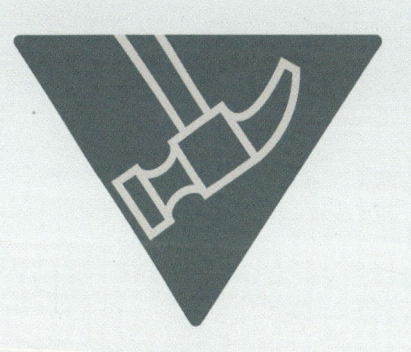

Step 4: Implement the Right Design That Will Most Improve Your Weakest Resource

Implement

im·ple·ment

/ˈimpləmənt/

verb

To execute a decision and/or design.

Credit: Kurt Stenberg

The dirty little secret of permaculture is that design is the easy part, it is implementation that kills you.

Like me, you have probably seen dozens of breathtaking permaculture property designs that *just felt right*. Maybe you have even created one yourself. But also like me, you can probably count on one hand the number of real-life permaculture properties you have visited that did not fill you with vicarious anxiety from all the half-baked and half-finished projects. Hopefully you haven't created one of these! But sadly, most of the permaculture properties I have visited seemed to be a kind of monster that consumes an increasing amount of time, money, and energy to keep alive. Worst of all, and unbeknownst to the property owners, as soon they run out of any one of their eight personal resources, the very thing that was meant to help them to survive and thrive will begin to cannibalize its own creators. The dirty little secret of permaculture is that design is the easy part, it is implementation that kills you.

I have almost been consumed by my own property, on several occasions! After these close calls, and seeing the same pattern with my students and clients, I began to realize that this monster is conjured into existence because of a failure to make good decisions about where to start or what is next, the fourth struggle of permaculture.

FIGURE 4.1. There is a very real threat of your property devouring you as your search for the blueberries that will provide your well-being.

After sharing this warning about implementation with a client named Errol, and my insights for how to avoid and, if necessary, stop this monster, he replied with a story about a unique kind of shotgun approach that perfectly captures the correct way to address the struggle of where to start or what is next.

During his time in British Columbia, Errol and his young family spent a lot of time picking wild blueberries. But because black bears and their cubs also love blueberries, he always carried a shotgun for self defence. This is a common practice, but Errol's method of loading with alternating birdshot and slugs in anticipation of being charged by a bear was quite unique. His reasoning was that a birdshot contained dozens of small metal projectiles designed to spread out in a wide pattern for hunting small fast-moving birds in flight, while a slug contains a single large projectile designed for hunting large stationary animals. By loading both these rounds in this order, he would have an almost certainty of actually hitting a charging bear without the time to take proper aim. And if the birdshot didn't scare the bear off, it would at least slow it down, providing time to collect his nerves and recalibrate his aim with a slug that would, if necessary, kill just about anything. If he missed this second shot (it is a charging bear after all), the next round was a birdshot guaranteed to at least slow or scare, followed by another slug intended to kill and so on.

Just like the story above, there is a very real threat of your property devouring you as you search for the blueberries that will provide your well-being. And there is an element of randomness and uncertainty as to when one of an unknown number of bears might attack, and how aggressively. Overcoming the struggle of where to start or what is next has as much to do with luck and intuition as it does with preparation and persistence. As such, the best approach to implementing your permaculture plan is to prepare with a process to stack luck and intuition in your favor, to be persistent with its use, and to be adaptive as things change. This process is to use both many small experiments spread as widely as possible, combined with a single concentrated effort with ongoing check-ins to make sure you are on track.

Step 4: Implement is all about our processes and tools that will help you to pick your blueberries and to be ready for those black bears with a birdshot and slug shotgun approach.

Vision without action is a daydream. Action without vision is a nightmare.

Japanese proverb

FIGURE 4.2. The birdshot, slug, buckshot, and bazooka are four analogies for how you may approach the implementation of your permaculture design.

By the end of this chapter you will learn:
- Why goal setting is nothing without goal execution.
- An innovative approach to loading a shotgun not for protection from bears but for building your permaculture property.
- The correct order of operations, in other words, workflow, to go about solving problems and actioning solutions.
- How to use a weekly planner to identify, organize, prioritize, and execute the goals related to your weakest link in your resources.
- How the four variables of pattern also apply to implementation.
- Why the fastest way to grow an apple tree might not be plant one as soon as possible.
- How to make good decisions and avoid bad or ugly ones.
- Our best advice when it comes to solving any problem.

What Is Your Birdshot?

Random experimentation is typically not something most permaculture practitioners need very much encouragement to start. In fact, most permaculture properties I visit or consult look like the laboratory of a mad scientist! But the birdshot approach to implementation is more than just taking shots in the dark. It is about taking aim at a known problem and firing dozens of very small experiments to see what, if anything, finds the mark. While taking shots in the dark leads to unintended casualties, a strategic birdshot leads to insights that prepare you to take aim for your second, much larger slug.

Before I discuss the birdshot solution to "where do I start or what's next," I will need to refresh your memory about the filestream and the SWOT diagnosis I introduced in Step 2: Diagnose. Let's review, and build upon these tools by starting with a pop quiz:

What is the answer to $(20 + (6 \times 3^2 - 14) \times (5 - 3)^2) / 60$?

Do you remember the first time your grade school teacher wrote something like this on the board and asked you to solve it? If you're like most people, you probably started on the left-hand side and performed each operator, carefully, from left to right. And when your teacher informed you your answer was incorrect, you were baffled. As it turns out there is a specific order that the problem had to be solved so as to reach the correct answer. In mathematics, the order of operations is encapsulated in the acronym PEDMAS, which represents the process: parentheses, exponents, division, multiplication, addition, subtraction.

Property implementation is no different. If you are getting the wrong answer while trying to solve problems on your property, it might not be because you don't have the right skills or you are using the wrong techniques—it's likely because you aren't approaching the steps of the problem in the correct order. The order of operations that I have found to work best for property implementation is the same as introduced earlier: 1. geography, 2. climate, 3. water, 4. access, 5. structures, 6. fencing, 7. flora, 8. fauna, 9. business, 10. technology, and 11. soil.

The reasoning for this is that the resource categories that are hardest and slowest for you to influence are generally at the top of the hierarchy, while those resources lower down are easier and faster for you to influence.[1] But also contributing to the order of operations is the relative influence that the categories have on each other. Those at the top of the list tend to have a relatively large influence on those below, while those at the bottom of the list tend

> **If you are getting the wrong answer while trying to solve problems on your property, it might not be because you don't have the right skills or you are using the wrong techniques—it's likely because you aren't approaching the steps of the problem in the correct order.**

> "Change your geography to fit your goals, or change your goals to fit your geography" is a motto I have said to many clients and students who bought the wrong property and were in the process of pounding a square peg into a round hole.

to have a relatively small influence on those above. As such, when implementing (or diagnosing or designing), I recommend working through these categories in an ordered workflow—from the first down through to the eleventh.

Geography is at the very top of the list, and ahead of climate, for several reasons. From the perspective of the property owner/permaculture designer, local geography is often slower to change and harder to influence than local climate (which is defined as the averages of day-to-day weather over thirty years). Furthermore, geography has a direct influence on your local climate, highlighting its importance relative to climate. For example, my neighbors on the other side of a small lake regularly report up to fifty percent differences in rainfall events. The disparity in between our farms is due to the presence of the lake, which in and of itself is a function of topography and geography. While it is true that humans can shape the land with bulldozers, geography comes before climate because in order to change your local climate, you would need to alter your local, regional, and even planetary scale geography.

On a more practical note, you can always construct a dam, build a greenhouse, or put up a windbreak to buffer weakness or threats from climate, but if you don't have a suitable geology, aspect, slope, or permission from your local regulator to do any of the above, you might as well move. "Change your geography to fit your goals, or change your goals to fit your geography" is a motto I have said to many clients and students who bought the wrong property and were in the process of pounding a square peg into a round hole.

Water is consideration number three in property design because all of the resources lower on the scale either need water or are harmed by its presence. Access is fourth because you can't manage any of your property resources if you can't access them. Winston Churchill's advice, "We shape our buildings, thereafter our buildings shape us,"[2] perfectly captures why structures are fifth. Next follows fencing (sixth), flora (seventh), and fauna (eight), listed in this order because no one would plant a garden before it was properly fenced or purchase a cow before they had pasture. Business is ninth because only once everything above is sorted out can you identify your production sweet spot. Technology comes in the tenth because you can have almost any tool imaginable delivered in weeks. But also because having technology second to last is a useful reminder that work is a failure in design, not a lack of equipment.

Last on the list comes soil. Due to the recent elevation to celebrity status, having soil as the eleventh (and last) category confuses many of my students, and has even angered some of my peers. Once during a panel discussion, I introduced these eleven categories only to have a fellow panelist, who happened to have a PhD in soil science, interrupt me to inform me that this model was nonsense. "Soils take an incredibly long time to form and therefore should be placed near the top" he argued. "Yes, that is partially true," I replied, and then described the two reasons that soil belongs at the bottom: You can lose a thousand years worth of soil in ten minutes via water or wind erosion if you don't manage your soil properly; and you can speed up the natural soil creation process and create a thousand years worth of soil in ten years if you manage your soil properly.

The misconception that improving soil takes centuries leads many people to place their kitchen garden a five-minute walk from the house because "there is really good soil there." The fact that they never use the garden because it is so inconvenient doesn't seem to factor into their design. The truth is that if you get the other factors in the earlier order of operations correct, your soil health will improve rapidly on its own.

With this workflow in mind, the answer to where to take aim first, where to start, can be found in the property resources SWOT diagnosis you created in practice 2.3. Simply start at the top with the things that are hardest and slowest to change and work your way down. However, oftentimes you will come across a strength, weakness, opportunity, or threat that you aren't certain how to leverage or mitigate. Because of all the problems of complexity, the root cause or the best solution could be just about anything, and it is all too easy to get bogged down with analysis paralysis that leads to nowhere, or burn out from the failure of too many Hail Mary attempts. So where do you start addressing these stubborn individual SWOTs? This is where the birdshot approach comes in. To identify some appropriate small-scale experiments to fire off at each specific SWOT, start by asking yourself this question, What are all the things I could do to address this that take less than one hour or cost the equivalent of my hourly wage?

With this workflow in mind, the answer to where to take aim first, where to start, can be found in the property resources SWOT diagnosis you created in practice 2.3.

> *You have to decide what your highest priorities are and have the courage—pleasantly, smilingly, unapologetically—to say no to other things. And the way you do that is by having a bigger "yes" burning inside.*
>
> Stephen R. Covey,[3] educator and author

Do you have some potential microclimates on a south-facing slope? You could take a variety of suckers, cuttings, or seeds from a diversity of existing plants to see what grows best there. Do you live in a climate prone to hailstorms? You could call all your neighbors to see if there was a market to graze damaged crops with livestock. Is the water well on your property slowly in decline but you have multiple ephemeral creeks? You could make one rock dam or gabion in your creek to increase water infiltration. Do you have minor erosion on your main driveway? You could place pieces of firewood at varying locations on your road so that traffic is directed around these speed bumps, and the act of driving would properly crown the road to get water to run off the road, not down it. Is the sump pump in your home running frequently? You could extend the length of your downspouts with some scrap pipe to get rainwater further away from your foundation. Are your fences beginning to clog up with brush? You could try intensive grazing around the fences to get your livestock to keep them clear. Are you having difficulty with certain species of plants or animals that should be doing well in your area? You could trade seeds or breeds with other permaculturists to experiment with dozens of varieties to see which perform best for your context. Is your permaculture business strapped for cash? You could analyze your chart of accounts to see what items currently cost you money and experiment with ways to cut expenses or better yet turn them into income generation. Is a piece of technology struggling to complete a task on the property? You could research other ways people have completed the same task, as far back as you can find records for. Are the soils in your garden in poor shape and not responding to other treatments? You could go for a walk with a shovel and dig as many holes as you can in varied ecosystems on or around your property for clues about what is missing.

All of these examples are things Takota or I have done or helped clients with when stuck on where to start on a specific SWOT. While very few of these experiments lead to an immediate solution, all of them provided valuable insights about where to take aim next with a more refined approach that lead to the eventual mitigation or leverage of that particular resource. The reason this approach works so well is that each of these small experiments will provide additional data points that are tailored to your context without costing more than a few dollars or a few hours of time. In short, this birdshot approach is an easy way to stack luck in your favor in a way that has a huge upside and no downside. Just as in diagnosis, once you gather enough of these data points, patterns will begin to emerge that will lead to the solution for the SWOT at hand as well as solutions for SWOTs not yet diagnosed! Once these small and broad birdshot experiments have shown their validity, you can scale them up into a good design, or if the problem persists and becomes a high enough priority, you will be better prepared to target it with a slug! More on that later.

It is also important to note that the same question—What are all the things I could do to address this SWOT that take less than one hour of time or cost the equivalent of my hourly wage?—works when you are stuck about where to start addressing SWOTs in your financial, social, living, material, experiential, intellectual, spiritual, or cultural resources. The only difference between where to start or what's next when implementing designs to address your personal resource categories is that you will have already created the hierarchy of most important to least important based on practice 1.3. While the order of your vision and values statements may change as you progress the development of your permaculture property, the order of operations is fixed and always stays the same.

Now that I have introduced the birdshot approach for minor SWOTs that get you stuck on "where to start or what's next," let us take a look at the slug approach for bringing down the major ones.

> p.s. $(20 + (6 \times 3^2 - 14) \times (5 - 3)^2) / 60 = 3$…If you don't follow the correct order of operations, you may not end up where you want to be (or get the correct answer).

What Is Your Slug?

Remember that good implementation, not good design, is your biggest barrier to success with building your permaculture property. As such, one of the most important things you can do is to use a disciplined approach, a process, that helps you execute your design as efficiently and effectively as possible, as well as adapt to changes as they come up. There are many frameworks out there for project management, especially if you look to the corporate world, and in

In short, this birdshot approach is an easy way to stack luck in your favor in a way that has a huge upside and no downside.

fact, the process I use myself and teach others is a simplified and slightly tweaked version of a framework[4] used by the executives of a number of million-dollar companies.

The basic premise is that the day-to-day operations should be kept to less than eighty percent of your available resources. These are the urgent tasks like to-do lists or birdshot experiments. If you are not disciplined, you'll end up caught in this whirlwind of activities and never make progress on the things that matter most. Bringing it back to our earlier analogy, if you want your slug shots to be effective, on an ongoing and iterative basis, you need to decide on your target and calibrate your sights. The 5SP Process uses the following four practices to do this: weakest link analysis, set a wildly important goal, describe lead measures, and make a weekly commitment. Let's go through each one.

Weakest Link Analysis

What is the biggest gap between the vision you want and the resources you have? Or what is causing the most suffering in your life right now? Or what is most limiting your well-being? Whatever the answer is, I call this your weakest link. Note that this is not a value judgement about your current inventory of what you have. It is a relative judgement about how your current resources are limiting the well-being described in the corresponding vision statement. Your weakest link then is the gap that is most limiting your overall well-being in the present. In the same way that a chain is only as strong as its weakest link, your well-being is only as strong as your weakest resources.

Wildly Important Goal

Once you have identified the weakest link in your personal resources or your property resources, you must turn it into a wildly important goal, called a WIG for short, that is achievable and gets you excited. I recommend that you focus on one WIG at a time because the more goals you have, the more divided your resources are and the less likely you are to achieve them. Unlike the qualitative and general vision statements written in Step 1: Clarify, when writing down your wildly important goal, it must be quantitative and very specific. Make it short, sweet, easily memorable, and hard for you or any of your partners to weasel out of! The formula "From X to Y by when"

> Remember that good implementation, not good design, is your biggest barrier to success with building your permaculture property.

works excellent. This WIG is your slug. It is the one shot, at present, that would take out that charging bear before it devours you.

For example, say that the biggest gap between your vision and resources is your concerns about your body weight. Your WIG for this weak link would be something like "I will go from 200 pounds to 150 pounds by next summer solstice." You might think that this is an unrelated WIG to building a permaculture property, but by far the most common weak links my students and clients identify when they get to this stage in the process are in living capital (personal health), social capital (relationship struggles), and spiritual capital (emotional trauma). The beautiful thing about this process is that it can help improve the health of your soil or your soul!

Another example, this time of a water resource weakest link, could be the "groundwater well flow rate is declining." Here your primary concern is related to a secure water supply. Following the formula, your WIG to address this issue could be stated as "from water scarcity to water abundance within one year" or "from declining well flow rate to increasing well flow rate within five years."

Lead Measures

The weakest link and the WIG are pretty straightforward, but the next step is to identify what specific actions you have control over that will most likely lead to achieving your WIG. Think of the goal of moving a giant rock. Lead measures can be thought of as the different kinds of pry bars you might try to use to move the large rock. Alternatively, you can think of these as the battles that help to win the war.

Consider that your weight is the number that shows up on the scale every time you step on it. As satisfying or disappointing as this might be, no matter how often or how accurately you measure this number, in itself it does not help you to lose weight. It only tells you what your weight is. Your actual weight is therefore not a lead measure. For the failing groundwater well example, knowing the flow rate volume of your well doesn't change the outcome either. Therefore, the well flow rate is also not a lead measure. Lead measures are the actions and efforts that you have direct control over and that influence or most likely lead to you achieving your WIG. They are your leverage points. For example, a lead measure for the weight-loss goal could be stated as "the number of calories consumed."

> **The beautiful thing about this process is that it can help improve the health of your soil or your soul!**

But it could also be: number of grams of carbohydrates I eat per day, number of hours per week at the gym, number of home-cooked meals per week, number of different diets experimented with, number of coaching sessions attended per month, number of hours a day of intermittent fasting. All of these lead measures predict weight loss, and they are directly influenceable by you.

For the failing groundwater well example, you can influence the number of liters of water infiltrated through swales and ponds on your land. "Number of liters of water infiltrated" is therefore a good lead measure. And sometimes (perhaps often), the problem you are trying to solve is complex, and it is really difficult to know what the best lead measure is. In that case, my favorite lead measure is simply "number of behaviors adapted, adopted, or eliminated." Now you are simply focused on influencing the number of different things you are trying to do, and watching carefully for which one is actually going to move the rock!

Note also that your lead measure shouldn't stipulate a specific quantity of an action. For instance, you wouldn't put "consume less than 50 grams a day of carbohydrates" or "infiltrate one thousand liters per day in my swales." This is because it is impossible to know exactly how much you need to push down on a particular pry bar to move a large rock, let alone how many pry bars are necessary or which ones are necessary. You can think of these as buckshots that accompany your slug! They are bigger and have more force behind them than your birdshot, but there are still several of them to improve your chances of hitting the target.

Weekly Commitment

Everyone knows the exhilarating experience of being on a roll and the paralyzing feeling of being stuck in a rut. Whether you are working with a partner towards a shared goal or individually, the best way to overcome inertia and create momentum towards your WIG is to set a weekly commitment to move that rock forward. These weekly commitments must be directly related to your lead measures, and typically they should address any critical issues or lead dominos. The question I like to ask myself is: What's one best action I can do this week that will have the biggest impact on my WIG? For weight lost, this might be: "This week I will clean out the house of all junk

---

Lead measures are the actions and efforts that you have direct control over and that influence or most likely lead to you achieving your wildly important goal.

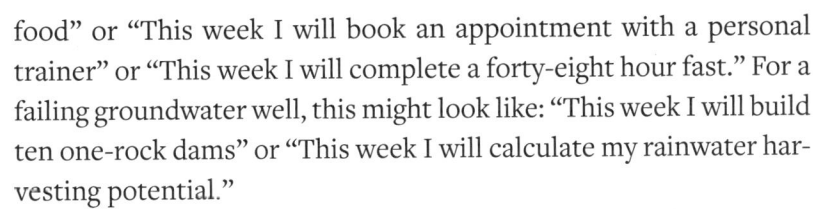

food" or "This week I will book an appointment with a personal trainer" or "This week I will complete a forty-eight hour fast." For a failing groundwater well, this might look like: "This week I will build ten one-rock dams" or "This week I will calculate my rainwater harvesting potential."

From experience, I have seen that most people are highly intuitive about what it is they need to do to improve their well-being, but they just feel hopeless about doing it. If you can get yourself and your partner(s) to pick up the biggest pry bar and just pull down once on it every week, it won't be long before you will start to see movement, and then you'll start to feel optimistic and energized about achieving your goal in no time.

Table 4.1 and Table 4.2 provide one example of a weakest link for each of the personal and property resources, and what a WIG, some lead measures, and weekly commitment could look like.

Pull the Trigger

Now that you are familiar with the birdshot, slug, and buckshot approach for implementation, it's time to start firing shots at your SWOT. Figure 4.3 shows an annotated version of the weekly planning template that Michelle, Takota, and I have used for years to help not only implement our permaculture designs but also better manage our personal lives. Every single Monday morning, I start with a blank sheet and indicate the date (or week number, or both) in the top right-hand corner of the page, see the letter A.

The top middle of the page, shown by the letter B, is a blank area for you to record and track your day-to-day tasks and any birdshot experiments you have on the go. The three columns help you break up activities into high, medium, and low priority or into other categories like personal, farm, work, or any other combination you see fit. As you finish tasks, you can cross them off; as new ones emerge, you can add them into appropriate locations. Whatever does not get finished you can transcribe, by hand, to a new template the following week. Trust me on this, you will find that your motivation to complete a task increases relative to the number of times you have to transcribe it!

The leftmost top column, shown by the letter C, is dedicated to your slug and buckshot. Here you record your weakest link, your

> **If you can get yourself and your partner(s) to pick up the biggest pry bar and just pull down once on it every week, it won't be long before you will start to see movement, and then you'll start to feel optimistic and energized about achieving your goal in no time.**

TABLE 4.1. Personal Resources Examples. One example of a weakest link for each of the personal resources, and what a WIG, some lead measures, and weekly commitments could look like.

	Weakest Link What is most limiting my well-being right now?	Wildly Important Goal From X to Y by when.	Lead Measure What do I have control over that most likely leads to achieving my WIG?	Commitment This week I will…
EXAMPLES	Credit card debt (Financial)	From $10k credit card debt to zero credit card debt by this January.	Number of completed courses on money management, number of professionals consulted, number of books read/videos watched about debt.	Attend the community course about credit card debt management.
	Personal health troubles (Living)	From terrible muscle and joint pain to pain-free life by my 40th birthday.	Number of health modalities consulted, number of people consulted with with similar issues/symptoms, number of minutes spent stretching every day.	Get a diagnosis from two different healing modalities.
	Poor relationship with spouse (Social)	From highly stressed communication to effortless communication by our next anniversary.	Number of counselling sessions, number of books read, number of respectful conversations.	Interview two relationship counselors.
	Too much stuff (Material)	From cluttered life to organized life by June 1 next year.	Number of emails/phone calls to friends or charities about donating stuff, number of garage sales.	Inventory everything I own on paper, dividing it into keep, donate, or recycle categories.
	No gardening skills (Experiential)	From zero gardening skills to growing 75% of all the vegetables I eat by next fall.	Number of conversations with other gardeners, number of workshops attended, number of books read.	Call 2 or 3 local farmers and book a tour of their operations.
	No content to bring customers to website (Intellectual)	From zero online content to one blog created each week by the next 6 months.	Number of hours of undistracted writing time.	Wake up and write for one hour every morning this week, before checking my email or getting distracted.
	Feeling hopeless (Spiritual)	From no desire to get out of bed to "can't wait to get up" by the next 3 months.	Number of hours spent volunteering, number of minutes in meditation, number of days off doing something I love.	Meditate every evening for 10 minutes before going to bed.
	No connection to community (Cultural)	From no community to thriving community by my 45th birthday.	Number of neighbors that I get to know, number of community groups that I join, number of potlucks that I host/attend.	Spend one hour researching and make a list of three local community groups that I would consider joining.

Step 4: Implement the Right Design

TABLE 4.2. Property Resources Examples. One example of a weakest link for each of the property resources, and what a WIG, some lead measures, and weekly commitments could look like.

	Weakest Link What is most limiting my well-being right now?	**Wildly Important Goal** From X to Y by when.	**Lead Measure** What do I have control over that most likely leads to achieving my WIG?	**Commitment** This week I will…
EXAMPLES	Hill above main road is sloughing (Geography)	From potential road washout to stable road by this coming spring thaw.	Number of hours learning patterns and principles of hill slippage, number of hours implementing tailored practices.	Call two bioengineering companies for advice.
	Prevalent hail belt (Climate)	From a property fragile to hail to a property antifragile to hail by next summer.	Number of hours in deep work, number of brainstorming sessions with accountability partner.	Create a detailed list of all the damages from hail on this property over the last ten years.
	Groundwater well flow rate is declining (Water)	From water scarcity to water abundance within one year.	Number of liters of water infiltrated.	Build ten one-rock dams.
	Always getting stuck on driveway (Access)	From always muddy driveway to high and dry access by next spring.	Number of earthmovers consulted about a new road.	Purchase contour maps.
	House roof is leaking (Structures)	From leaking roof to 100% dry home by this coming rainy season.	Number of contractors contacted for estimate.	Call all the people I know who have done renovations and get recommendations.
	Poor perimeter fencing (Fencing)	From animals escaping every week to never having a rodeo again by September 1 this year.	Number of hours spent researching fencing design, number of hours spent implementing that design.	Advertise for a part-time helper this summer at the local high schools.
	High mortality on tree seedlings planted this year for our shelterbelt (Flora)	From 50% tree seedling mortality to less than 5% mortality for next year's tree planting.	Number of hours learning patterns and principles of tree seedling health, number of tailored practices implemented.	Contact the nursery who supplied the seedlings, share my experience, and ask for advice.
	Prevalent mastitis in dairy cows (Fauna)	From spending hours treating mastitis symptoms to healthy happy cows that never get sick.	Numbers of behaviors and systems adapted, adopted, and eliminated.	Get nutrient testing on their feed.

TABLE 4.2. (cont'd.) Property Resources Examples. One example of a weakest link for each of the property resources, and what a WIG, some lead measures, and weekly commitments could look like.

	Weakest Link What is most limiting my well-being right now?	**Wildly Important Goal** From X to Y by when.	**Lead Measure** What do I have control over that most likely leads to achieving my WIG?	**Commitment** This week I will…
EXAMPLES	Poor farm profitability (Business)	From $25K net farm profit to $50K net farm profit by next fall.	Number of new products created, number of expenses cut.	Interview two business coaches.
	Spending too much time maintaining machinery (Technology)	From high dependency on machinery to minimal dependency on machinery.	Number of machinery-dependent tasks designed away, number of hours in deep work.	Make a list of each piece of equipment and its associated needs and yields.
	Poor soil health is limiting productivity (Soil)	From soil like bricks to soil like chocolate cake by the next five years.	Number of hours learning patterns and principles of soil health, number of tailored practices implemented.	Complete a visual soil health assessment for all different soil types on the property.

> Once you've recorded these, it's important to verbally share your commitment with your property partner or snap a quick photo and send it to your accountability partner.

wildly important goal, lead measures, and an actionable commitment for the week. Once you've recorded these, it's important to verbally share your commitment with your property partner or snap a quick photo and send it to your accountability partner.

A daily timetable is included on the bottom third of the page, shown by the letter D. On Monday morning, I'll actually transcribe all weekly appointments from my digital calendar into my analog one, and that brings stability and sanity into my life. I also use this area to then reflect on my weekly commitment and my lead measures, and schedule in the necessary time to execute.

Depending on your own current WIG and lead measures, it's also easy to use the weekly planning page, particularly the daily timetable area, as a scoreboard as the week progresses. In the weight loss example, this could be recording your weight each day beside the weekday name, combined with a big smiley icon along with the actual number of carbohydrates you consumed, or the actual number of liters of water you infiltrated (i.e., the lead measure you are currently targeting). The goal here is to motivate yourself by providing

Step 4: Implement the Right Design

A WEEK:

C

WEAKEST LINK
What is most limiting my well-being right now?

WILDLY IMPORTANT GOAL
From X to Y by when.

LEAD MEASURES
What do I have control over that most likely leads to achieving my WIG?

COMMITMENT
This week I will...

B

	MONDAY	TUESDAY	WEDNESDAY	THURSDAY	FRIDAY	SATURDAY	SUNDAY
MORNING							
7 AM							
8 AM							
9 AM							
10 AM							
11 AM							
12 PM				**D**			
1 PM							
2 PM							
3 PM							
4 PM							
5 PM							
EVENING							

FIGURE 4.3. Every week my co-authors and I use a weekly planning page, just like this one, to manage both our permaculture properties and our personal lives.

clear indicators of progress or regress, and to track yourself in a way that can reveal patterns about how you may increase the leverage of your lead measure, or drop lead measures that aren't doing anything. For example, maybe after several weeks of seeing a large drop in your weight after the weekend, you might realize that on the days after your weekend fast, you are seeing the greatest reduction in weight loss, and that may be due to the inflammation and associated water weight from something in your diet during your busy work week. And maybe instead of "number of liters of water infiltrated," you might decide to shift your focus to a lead measure such as "number of practices adopted that are known to increase soil carbon." I'll talk more about the relationship between soil carbon and the water cycle in the next chapter; the point here being that you want to focus on things that you have direct control over, and when stuck, try manipulating the scale, timing, placement, and form independently to look for clues of what might work.

Of course, you don't have to use the template exactly as shown, and you can modify the steps to work for you, especially if you already know and use a different framework for project management. The big idea is to make sure you don't get fully caught up in your day-to-day activities and do set yourself up to be disciplined to focus some effort, every week, on what will most influence achieving your goal of building out the property, and the life, of your dreams. Defining your weakest link, setting one wildly important goal at a time, acting on lead measures and holding yourself accountable through a weekly commitment shared with your partner are disciplines that are simple, yet effective and have worked very well for myself and many others.

Good, Bad, and Ugly Decisions

Back in Step 1: Clarify, I discussed the individual, community, and global implications of the struggle to figure out *What to do* and how clarifying *What you want*, *What you have*, and *What is right* can help. But even if we are all on the same page, figuring out *What to do* can still be a difficult task when it comes down to the details. This in turn leads people stuck not knowing where to start, or what is next. Remember also that in Step 3: Design, I defined three types of design: good, bad, and ugly. Similarly, I'm proposing that there are three types of decisions. Decisions that are unethical are just plain

---

The big idea is to make sure you don't get fully caught up in your day-to-day activities and do set yourself up to be disciplined to focus some effort, every week, on what will most influence achieving your goal of building out the property, and the life, of your dreams.

TAKOTA'S STORY | The Bazooka Approach

In an article titled "Phases of Abundance," Bill Mollison[5] laid out the following five-year plan for how to take $800 worth of plants composed of hundreds of species and dozens of varieties and turn it into a thriving food production system:

- Season 1. Establishment of an abundant richness of species for trial.
- Season 2. An abundant source of propagation material is produced.
- Season 3. Several species are sufficiently numerous to provide an abundance of yield.
- Season 4. Yield is excellent, propagation material "unlimited," and slower species start to produce. Absolute abundance is achieved.
- Season 5 on. Abundant yield from as many as 30 perennials and the same number of annuals is achieved, and can be made to persist for as long as is needed.

I now recognize this as a perfect example of a birdshot approach for plant selection; however, when I read it early in my permaculture journey, I did not appreciate its brilliance and instead opted for the "bazooka approach" when it came to implementing my first food forest that was supposed to address my weak link of food security.

My logic was based on the proverb "The best time to plant a tree was twenty years ago; the second-best time to plant one is today." I wanted to have an abundance of fruit. It takes many years for perennials to grow and produce, particularly in cold climates. Therefore, the sooner, the larger, and the more fruit trees I plant, the sooner and the greater the abundance I would have. Boy was I wrong.

That fall, after interviewing a few nurseries about the two to three "best" varieties for our area, I planted seventy apples, plums, apricots, and pears taller than myself worth about $7,000. So confident was I that I could achieve abundance in one season instead of five, I even started to advertise that our farm would have fruit for sale by next fall. I assure you this was not a "slug," it was a "bazooka." And as you will see, putting too many resources towards solving a SWOT has the risk of backfire.

That first season, all but three of those seventy fruit trees had died. In diagnosing the cause of their death, I discovered that all of the trees were root-bound from being kept in pots too long. Justifying that this failure was due solely to an incompetent, or more likely dishonest, tree nursery, I decided to try again, but this time I planned to pay a little more attention to Bill's birdshot advice by increasing the diversity of my species and varieties. In addition to apples, pears, plums, and apricots, I planted oaks, black walnuts, hazelnuts, cherries, cherry plums, sand cherries, raspberries, haskaps, currants, gooseberries, saskatoon, seabuckthorn, buffalo berry, Russian olive, caragana, strawberries, blueberries, and highbush cranberries. I planted up to six varieties of each. I also opted for smaller stock to minimize the chance of root binding and to keep costs per plant low.

However, hoping to make up for the lost time of my first season, I doubled down and purchased 5,500 plants for planting in the next. This round of bazooka ammunition cost over $10,000. Not surprisingly, this second Hail Mary also partially backfired, but this time it was entirely my fault, and no fault of the tree nursery. Despite my best efforts, by the end of that second season, about fifty percent of species and varieties were dead or dying from too much or not enough of what I call the 4Ws of tree planting: wind, water, wildlife, and weeds.

As expensive and demoralizing as these failed experiments were, I was still able to gather some valuable insights that have given me a much greater respect for the birdshot approach laid out by Bill.

FIGURE 4.4. Takota in his six-year-old forest garden enjoying his first crop of apples.

The first insight was about the importance of scale, timing, placement, and form when it comes to implementation. Planting a one-acre food forest ten steps from your house is much different than planting a six-acre forest garden one thousand steps away. I found out the hard way that there is sound reasoning behind the concept of permaculture zones. The reason you put the things that need you or that you need the closest to you is that every extra step you take away from your home, the surface area you have to manage increases exponentially. This can be easily seen in the formula for the area of a circle: πr^2. Factor in the three-dimensional world and the fourth dimension of time, and it is easy to push the birdshot approach beyond your capacity to observe the insights from your experiments, try as hard as you might.

Another insight came from a totally unintended birdshot experiment. As a thank-you to my sisters for helping me plant those 5,500 trees, I gave them a few of each of the same species and varieties. To my astonishment, by the fifth season when I was finally starting to get decent yields, just like Bill had promised, their same perennials were twice the size of mine and with twice the production. This is likely because they had fewer plants and were able to give each one much greater care, but also because they had much better microclimates. As it turns out, it is not the sooner the larger and the more fruit trees you plant, the sooner and the greater the abundance you will get. It is the sooner you create the conditions for the apple tree to thrive, the sooner you will get apples.

Proper birdshot experiments done at the right scale, right time, right place, and the right form really do provide you with valuable insights. By observing which species did well despite my inability to properly care for them, I have been able to plant up to 2,000 trees per year with very little effort and a ninety-five percent success rate. But sadly, I could have had the same insights with a fraction of the cost, and a great deal more enjoyment.

ugly, decisions that are both ethical and right for you and your land are *good*, and decisions that may be ethical but perhaps wrong for you or your land are bad. Your goal, when developing and building your permaculture property, should be to consistently make as many *good* decisions as possible, minimize the number of bad decisions, and as much as possible, avoid ugly decisions.

So, let me ask you this: *What is your process for making good decisions?*

When I ask this question in class or on a consult, most people, after a long reflective pause, typically say, "I usually just go with my gut" or "I don't really have one." However, only using your intuition or having no evaluation process at all are both one-way tickets to burnout. Before Michelle and I started using a process, decisions took a long time and we were never sure if we were making the best choice. We half-hazardly used tools like pros-and-cons lists and

What is your process for making good decisions?

THE GOOD DECISION WORKSHEET
(How to avoid bad and ugly decisions)

THE FIVE-STEP PERMACULTURE PROCESS

▶ PASS CONTINUE TO NEXT STEP
▶ FAIL RESEARCH AND/OR REVISE - RETURN TO START

START

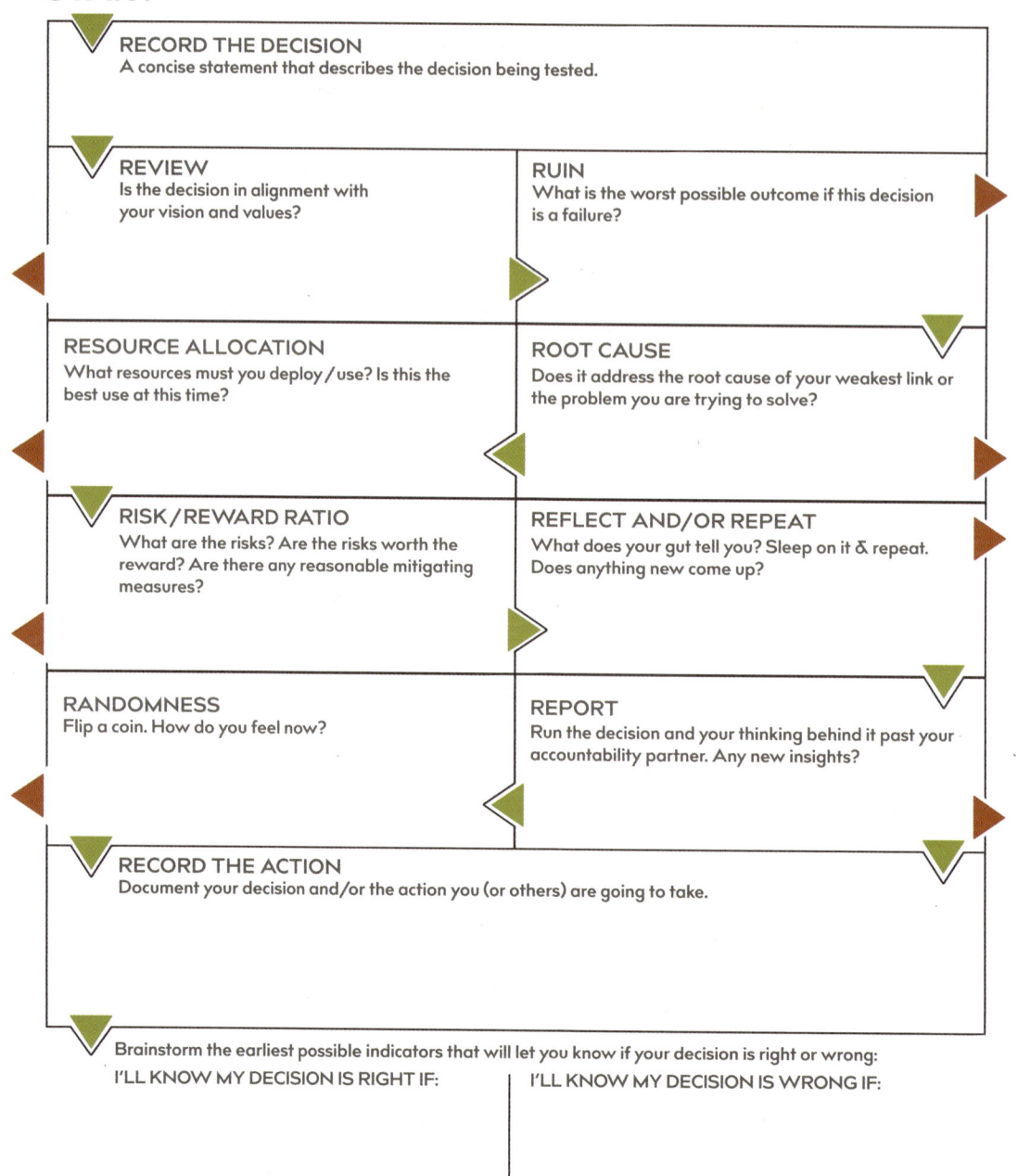

RECORD THE DECISION
A concise statement that describes the decision being tested.

REVIEW
Is the decision in alignment with your vision and values?

RUIN
What is the worst possible outcome if this decision is a failure?

RESOURCE ALLOCATION
What resources must you deploy/use? Is this the best use at this time?

ROOT CAUSE
Does it address the root cause of your weakest link or the problem you are trying to solve?

RISK/REWARD RATIO
What are the risks? Are the risks worth the reward? Are there any reasonable mitigating measures?

REFLECT AND/OR REPEAT
What does your gut tell you? Sleep on it & repeat. Does anything new come up?

RANDOMNESS
Flip a coin. How do you feel now?

REPORT
Run the decision and your thinking behind it past your accountability partner. Any new insights?

RECORD THE ACTION
Document your decision and/or the action you (or others) are going to take.

Brainstorm the earliest possible indicators that will let you know if your decision is right or wrong:

I'LL KNOW MY DECISION IS RIGHT IF: I'LL KNOW MY DECISION IS WRONG IF:

FIGURE 4.5. Whenever you need to make a big decision, and/or whenever you want to avoid making bad and ugly decisions, the Good Decision Worksheet is just that, a one-page worksheet that you pull out and use as a guide.

financial comparisons, but it always felt clunky. I remember once trying to make a decision about a used car, and it took over a year for us to come to a conclusion. This is fine if you only have one decision to make at any given time, but when you are trying to develop a property, you need a quick framework that gets all stakeholders quickly onboard and allows you to make decisions you can trust. When we first discovered, and started using, a process for making decisions as taught by Allan Savory,[6] we were blown away with a faster ability to make big decisions confidently. After years of using that framework, we started making tweaks, and with Takota's input, the decision-making framework we now teach and use evolved into what we call *The Good Decision Worksheet*.

The Good Decision Worksheet

Whenever you need to make a big decision, and/or whenever you want to avoid making bad and ugly decisions, the Good Decision Worksheet is just that, a one-page worksheet that you pull out and use as a guide. It is shown in Figure 4.5.

Simply start at the top of the page and follow the arrows while working through all of the cues in the boxes. A green arrow represents a "pass," and that means that you get to move on to the next box. However, if you can't confidently pass the cue, or if your answer is "maybe," you are prompted to follow the red arrow, which is labeled "research and/or revise" and takes you back to the beginning. This means that you should either put away the worksheet and go do more research or revise the decision that is being made while starting at the top again.

If you pass all of the boxes and get to the bottom of the worksheet, you are very likely making a good decision, or at least the best decision you can with the information you have at hand. Let's go through each box and the corresponding cues in more detail.

Record the Decision

Write a concise statement that describes the decision being tested. This is valuable because it is not uncommon for the stakeholders to be unclear about what they are all deciding on. As an example, Michelle and I think very differently and from time to time, when making a decision, we will think that we are both talking about the

> **When you are trying to develop a property, you need a quick framework that gets all stakeholders quickly onboard and allows you to make decisions you can trust.**

same thing only to find out that the decision we are trying to make is framed differently by each of us. Take time to make sure all stakeholders are on the same page. What is it you are deciding upon? How much money are you looking to spend? What resources are you about to deploy? Be as specific, yet concise, as possible. Also, occasionally you will have multiple options that you are considering. If this is the case, you will want to be clear about the option that is on the table for consideration.

Review

Review your vision and values one-pager: is this decision in alignment with what you want and what is right? If the answer is no, then you have a few options. You could follow the red arrow and revise your decision, then try again. Or alternatively, you could simply admit to yourself that you are going to do something that is ugly. If that's the case, the worksheet stops here. Yes, life is nuanced and complex, and ugly decisions do happen. But all is not lost. By simply admitting to yourself, on paper, that your decision doesn't meet your own ethics, you are far more likely to start making more and more good decisions—and subsequently getting to the end of the worksheet more often!

If the action or decision is in alignment with your vision and values, then it passes this test and you continue on.

Ruin

What is the worst possible outcome if this decision is a failure? Is there any chance it will lead to suffering that would be difficult or impossible to repair?

There is a distinct difference between risk and ruin, and if taking this action has the potential to cause ruin, you should probably not take it. "Ruin" in the context of this framework can be defined as any amount of suffering beyond a level that you or the people or systems you are deciding on behalf of can tolerate. While most decisions will carry risk, we should aim to avoid taking decisions that can be ruinous.

If, while running this test, you determine that this action needs to be taken in spite of the risk of ruin, the next question you should ask is, are there steps you can take to mitigate this outcome?

Root Cause

Does this decision address the root cause of your current weakest link or the problem you are trying to address?

In this question, you want to establish if the action you intend on taking is going to address the root cause of the problem. The majority of human solutions, especially in agriculture, are nothing more than techniques designed to address symptoms. Weed killers, fungicides, pesticides, and insecticides are Band-Aids designed to get rid of the undesirable organism, which is there because you may have created the conditions to favor it. Why is it there in the first place and what can you change to bring the system back into balance? When taking an action, make sure that you are confident that it is getting to the root cause, not just a symptom.

Resource Allocation

List the specific resources you will be using to action this decision. Make sure to consider all eight categories of your personal resources and possibly your property resources as well. Is this the best use of your, or your property, resources at this time? Could you use the same resources in a different way to achieve better results?

Risk-Reward Ratio

What are the risks associated with taking this decision? Are the risks worth the reward? What are the pros? What are the cons? Are you comfortable with the risk-reward ratio? If there are risks associated with this decision, what mitigation strategies can be employed to manage the risk?

Go through your resources and determine if you are creating any undue risks. Don't forget to look at some of the harder to quantify risks like social and living resources. A social risk might include a damaged relationship or reputation with specific people, while a risk to your living resources could have an effect on the amount of free time you have or compromise biological systems you depend on.

Reflect and/or Repeat

Do a gut check on all the work you have done so far. How does it feel? You may also choose to sleep on the decision for at least one night. After this period of reflection and rest, repeat all the cues from the start. Did anything new come up? Do you still want to make this decision?

Report

Report your proposed decision and your reasoning to your accountability partner—that is, someone who has a good understanding of your vision, values, and resources and is not directly influenced in the outcome of the decision. You need someone who can call a spade a spade. Through coaching sessions, over and over again, I've seen clients rationalize a decision not because it is necessarily good but because they've already made up their mind before even starting the worksheet. An honest and unbiased friend can challenge you and help you see how powerful your cognitive biases really are. Takota and I use each other as sounding boards for most big decisions to this day. The insights we have gained from each other's clarity of distance has been both humbling and invaluable.

Randomness

Sometimes, after going through all of the cues, you are still left feeling unsure. In that case, use randomness to make the decision for you. Open any book to a random page and read the first paragraph to see if it has any significance. Flip a coin: heads you do it, tails you don't. The beauty of this is that, oftentimes, you will feel instant relief that "fate" really did choose the option you wanted. Or if you feel disappointed at the thought of having to do "that thing," then you can ignore chance and choose what you had wanted all along.

Record the Action

Hurray! You've made it to the bottom of the worksheet, and you should feel fairly confident that your decision is good. Record any differences in the final decision from the original one at the top of the page, and it's also a great idea to assign responsibility if it makes sense to do so.

Right or Wrong

But you are not quite done. As humility is an essential part of managing complex systems, you should take the time to consider the consequences of a bad decision, what symptoms to look for, and how to pivot if this is the case. When you take the time to premedi-

> **When you think that you may have an answer to the question *Where to start?* or *What's next?*, a decision-making framework, such as the Good Decision Worksheet, is an indispensable tool.**

tate what a bad outcome could look like, it opens your perception up so that you are paying attention to the signals that may indicate a problem, and you are then far more likely to course correct in a timely fashion. Just think back to the gorilla in the room.

So take a minute and brainstorm the earliest possible indicators that you might perceive will let you know if your decision is right, or if it is wrong, and record these in the boxes at the bottom of the page.

It's a good idea to file your worksheet somewhere you can find it later if necessary. Several times we have made a decision only to second-guess ourselves a few months later. When I pulled the sheet out, I was able to remind myself why we made that choice, and any concerns fade.

More than anything else, I've found that the framework has actually helped us to say no more often, even to good ideas. There will always be more good ideas than there is capacity to execute, and it is better to do few things well than many things poorly. And if you have an unlimited firehose of ideas (like I do), this might be what your partner likes best about the framework (just like Michelle does).

When you think that you may have an answer to the question *Where to start?* or *What's next?*, a decision-making framework, such as the Good Decision Worksheet, is an indispensable tool. Just like any new tool, it can take a little getting used to using, but I assure you that, with a little practice, you'll be making better decisions, with less conflict and with increased peace of mind, faster than you ever have before.

Practices for Step 4: Implement

Practice 4.1: Plan Your Week, Weakest Link Analysis, Set a Wildly Important Goal, Describe Lead Measures, and Make a Commitment.

Download Template 4.1: Weekly Planning Page, or build your own based on the template in Figure 4.3.

On Monday mornings (or whatever day works best for you), fill in your weakest link, your wildly important goal, and your lead measures. Then, make a commitment—something that you will get

TAKOTA'S STORY | My Best Advice for Solving Any Problem

One of my instructors at trade school used to tell our class, "You can solve any problem in the world with three phone calls or less." This problem-solving hack is based on the immense power of social networking, and the ability to connect with an exponentially larger number of people through our immediate contacts. Although perhaps not scientifically proven, there is a general belief that we are all connected through no more than six degrees of separation.[7]

Regardless of the actual number of connections, the practical application of this idea is that any problem you encounter can be answered by first calling the most knowledgeable person you personally know about that particular problem and ask them the following: *"I am having trouble with X, and I know it's in your area of expertise. Do you have five minutes for me to ask you a few questions?"*

If your friend doesn't know the answer, the follow-up question that starts you down the networking path is: *"Is there anyone else you know personally I could talk to about X?"* You then call that next person, mention the name of the last person you talked to, and then ask them the same question, *"I am having trouble with X, and I know it's in your area of expertise. Do you have five minutes for me to ask you a few questions?"* Once again, if they don't know, ask, *"Is there anyone else you know personally I could talk to about X?"*

I use this approach all the time, and I have rarely had to contact more than three people before I get a direct answer for the problem I am trying to solve, or at least find the person I needed to hire to help me to fix it. The value of this problem-solving hack alone makes it worthwhile to create the social resource inventory discussed in practice 1.2 so that you can start to map out all the people in your life and the areas of expertise they have, along with their phone numbers.

In the age of "just Google it," instant messaging, and email, picking up the phone may seem antiquated. But once you try this a few times, you will find that there is no better or quicker way to receive tailor-made advice. You may also think that this is bothersome, but just remember the last time a friend, or a friend of a friend, called you for advice in your area of expertise. Everyone loves the feeling of being needed and of contributing to the larger community we are all a part of.

done this week and that will progress you towards achieving your WIG. Share your commitment with your partner, or alternatively with your accountability partner. This could be as simple as taking a photo of your weekly commitment and sending them a text message.

Copy any leftover to-dos from last week's planning page onto your to-do list for this week. After you transcribe your meetings, work, or other scheduled activities into your daily timetable/planner, schedule time to progress your commitment. Try using the page as a scoreboard and record lead measures for the day, or for the week, and compare these against any changes in outcomes. Are you closing the gap towards achieving your goal? If not, consider iterating your lead measure(s). Remember that one way you can iterate your lead measure is to change some or all of the following: scale, timing, placement, and/or form.

Practice 4.2: Make Good Decisions, and Avoid Bad and Ugly Ones.

Download Template 4.2: The Good Decision Worksheet, or build your own based on the template in Figure 4.5.

Sit down with your partner(s) and agree that you will use the worksheet for every decision that meets some some minimum criteria. For instance, commit to each other that any expenditure above $500 or five hours of time should be tested using the worksheet, but the exact number will be specific to your circumstances. If you have no partners, simply decide on what works for you.

Practice 4.3: What Are Your Birdshots?

Pull out your personal resources and property resources SWOT diagnoses from practice 2.3. Start at the top with the things that are hardest and slowest to change and work your way down. As you do, ask yourself *What are all the things I could do to address this SWOT that take less than one hour of time or cost the equivalent of my hourly wage?* Start incorporating ideas and experiments into your weekly planning page as time permits.

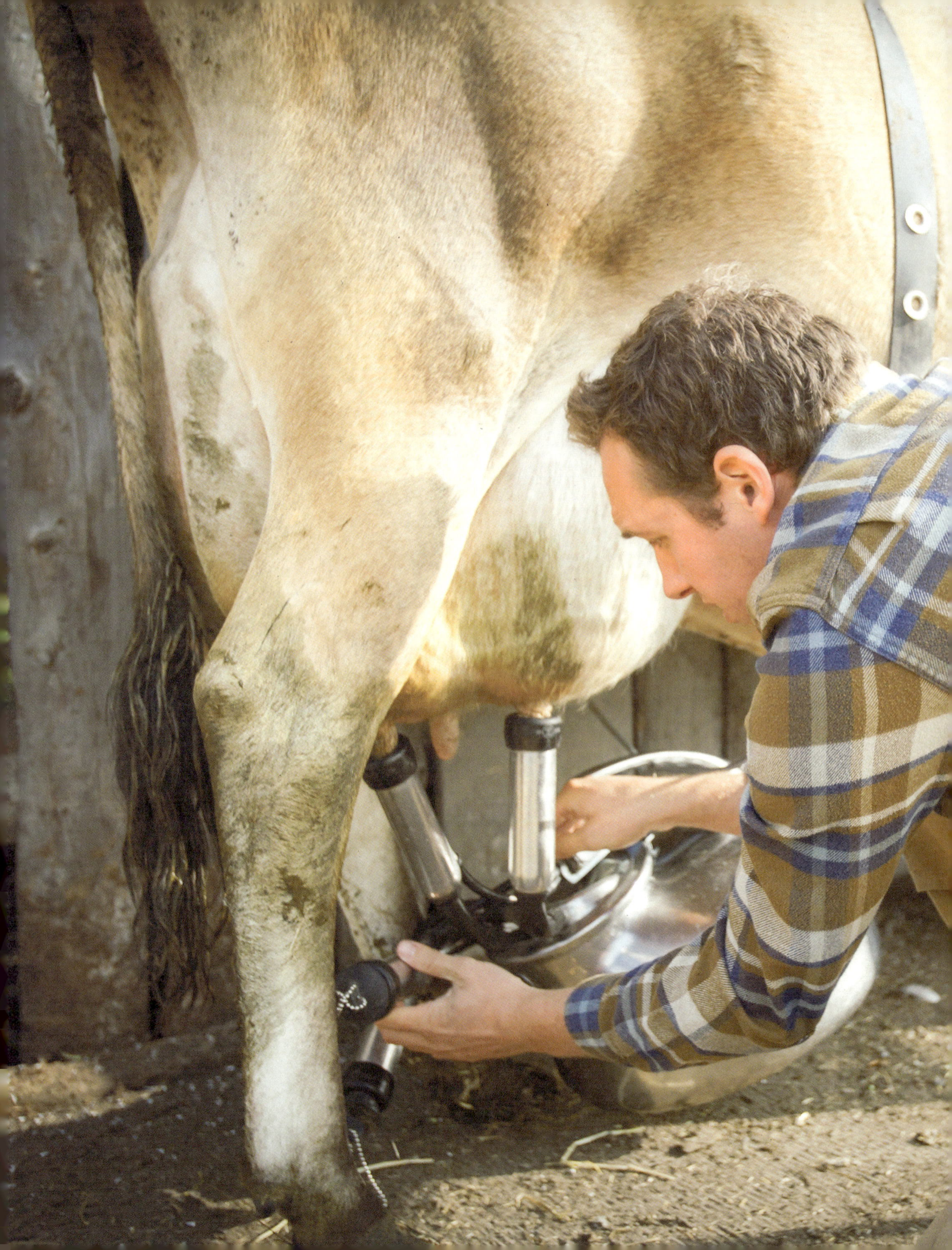

Step 5: Monitor Your Resources for Indicators of Well-being or Suffering

Monitor

mon·i·tor

ˈmä-nə-tər

verb

To observe, record, or check frequently for a special purpose.

Credit: Kurt Stenberg

> *We can find out about many things, both living and inorganic, by timing, measuring, and observing them; enough to make calendars, computers, clocks, meters, and rulers but not ever enough to understand the complex actions in even a simple living system.*
>
> Bill Mollison[1]

You can only manage what you measure. This oft-quoted maxim and its numerous spinoffs are as variously attributed as they are wrong. While measurement is essential when it comes to the specific lead measures I discussed in Step 4, measuring the random, emergent, and self-organizing aspects of complex systems is unnecessary, counterproductive, and in many cases, impossible.

Yet measurement remains a growing obsession when it comes to managing land. For example, "precision agriculture" uses satellites, drones, autonomous robots, machine learning, real-time sensors, and wireless transmitters to measure crop yield, normalized difference vegetation index, terrain features/topography, organic matter content, moisture levels, nitrogen levels, chlorophyll levels, pH, electrical conductivity, magnesium, potassium, and many more other variables. The goal of these precise measurements is to input the data into a software program for whole farm management so that seed, fertilizer, water, biocides, and other inputs can be "optimized" while outputs are increased.

In terms of the unnecessary and impossible aspect of measurement, think back to Figure 2.7 and remember that there is a world of difference between precision and accuracy! Even if it was possible to be one hundred percent precise in your measurement of the pH level of the soil or the number of beetles per plant, you could still be one hundred percent inaccurate in your assessment about what those measurements indicate and what to do about them. In other words, there is no right way to do the wrong thing.

All this precision ag technology has taken decades and billions of dollars to develop. Yet the zealous proponents of precision agriculture have failed to realize something that makes their endeavour a fool's errand. This is that the embodied energy of their crops, when you include the petroleum products required to measure, control, fertilize, till, seed, harrow, spray, harvest, transport, and process, is at least ten times more than the equivalent number of calories derived from their consumption.[2] That means that ten calories of petroleum energy are used to produce one calorie of food. I could also go on about other issues, including peak phosphorus, the growing dead zone in the Gulf of Mexico, ongoing depletion of our aquifers, erosion of precious topsoil, desertification due to irrigation and tillage practices, loss of nutrient density of our food, and species extinction, but at the end of the day, it is a matter of when, not if, the petroleum energy used to manage these systems will run out. For all these reasons, measuring is counterproductive.

As far as I can tell, precision agriculture and all the other ideologies obsessed with measuring complex systems are an attempt to achieve what Wendell Berry proposed as "the two great aims of industrialism—replacement of people by technology and concentration of wealth into the hands of a small plutocracy."[3]

As a result of the intuitive futility of measuring complex systems, many permaculturists tend to adopt a come-what-may approach to managing their permaculture properties. However, this is a false dichotomy that either reduces progression towards well-being or leads to something worse: after struggling with *What to do*, *Where to find*, *How to connect*, and *Where to start*, it is only a matter of time before you arrive at the fifth and final struggle of permaculture: *When will it end?*

The solution to the financial, environmental, and emotional bankruptcy is to monitor, not to measure. You might think this is a distinction that makes no difference. But the former cautiously avoids the myth of objectivity and never mistakes the what for the why! To measure something implies a passive, hands-off approach that leads to something final. To monitor something implies an ongoing flux of change that requires constant attention and, of course, management. This is why I prefer the phrase *only monitor what you can manage because you can only manage what you can monitor*. There are an endless number of variables to measure and an endless number of ways to measure them. Yes, although accurately measuring things like your bank account, rainfall, and yields will be valuable to aid your management decisions, there is a tendency to become seduced by the allure of measurement more than what is a common-sense approach. If you really want to avoid burnout, then you need to forget numbers and focus instead on something more nuanced: monitoring for suffering and well-being.

In this chapter, for the final step in the 5SP Process, I will share how to monitor your personal and property resources to optimize well-being and minimize suffering. You'll learn:

▶ The two most important criteria to be monitoring.

If you really want to avoid burnout, then you need to forget numbers and focus instead on something more nuanced: monitoring for suffering and well-being

- How to use journaling and planning tools to keep yourself on track towards your visions and values.
- How, and why, you should monitor the four ecosystem processes, and what these are.

The Push and Pull of Life

I have a riddle for you: What is the one thing never found in a healthy ecosystem?

If you are stumped, I'll provide a clue in the form of some Greek mythology about an ancient Greek king named Sisyphus who was considered by many, including himself, to be the smartest man to have ever lived. The problem was that Sisyphus used his intelligence towards avaricious and deceitful ends at the expense of his citizens, travellers to his lands, his own family, and even the gods.

The height of his treachery came at the end of his life when he fooled Hades, god of the underworld, into imprisoning himself in his own chains that were originally meant to drag Sisyphus to the afterlife. With the god of death imprisoned, no human could now die. Sisyphus was mighty pleased with this coup de grâce until he realized that nothing could die, including animals. Infuriated that humans had stopped making animal sacrifices, the remaining gods made life unbearable. To quell his own suffering, Sisyphus begrudgingly welcomed his death and released Hades. But the mortal king still had not learned his lesson. In his dying wish, he told his wife to defile his corpse in the city streets. This scheme was an attempt to find sympathy from the gods so they would allow him to return to the mortal world out of pity for how poorly his wife had treated him. This plot too was soon uncovered by the gods, and as punishment for this and his many other unethical deeds, Zeus, king of the gods, sentenced Sisyphus to carry a boulder up a mountain. I'm sure he couldn't believe his luck with this slap on the wrist until he realized the boulder was enchanted to forever roll back to level ground before the summit could be reached. And so, Sisyphus was trapped for eternity in a laborious, futile, impossible, *Sisyphean* task: pushing a rock up a hill.

So what does this ancient myth have to do with my riddle about healthy ecosystems, much less building a perma-

---

Zeus, who guided mortals to be wise, has established his fixed law — wisdom comes through suffering. Trouble, with its memories of pain, drips in our hearts as we try to sleep, so men against their will learn to practice moderation.

— Aeschylus,[4] Greek playwright, 525 BC–456 BC

culture property? First let me answer the riddle. The one thing never found in a healthy ecosystem is excessive suffering, not for others, nor for yourself. Of course, the operative words here are "excessive" and "healthy." Some suffering is omnipresent in life. And a good thing too! Just think how short the life of any creature would be without the motivating pain of hunger, thirst, cold, or fire.

Previously, I have defined well-being as the ongoing pursuit of the optimal expression of all your resources in an ethical manner. It is a state of being that translates to successfully actualizing your potential, expressing the very best within yourself, and encouraging all other organisms to do the same. It could also be described as the equilibrium point reached by an animal or organism through its niched work of digging, scratching, browsing, pecking, playing, mating, and rearing young.

This is why being in nature is so universally described as calming, blissful, and even medically therapeutic.[5] Being trapped in a dense smog during rush-hour traffic, or pushing a rock up a hill, is the exact opposite. All of these are ecosystems, only one of them is healthy. In the myth of Sisyphus, I see Zeus's use of suffering as a metaphor to teach Sisyphus the same lesson that the invisible hand of complexity "teaches" organisms in a healthy ecosystem. And that lesson is simple: If you spend any time observing healthy ecosystems, particularly wild animals, it is easy to see that this suffering and its opposite well-being are literally the push-and-the-pull force that guides the actions and evolution of all life.

Our present downward spiral bears many similarities to the myth of Sisyphus when we consider the following. In our self-righteousness, we believe ourselves to be the smartest creature that has ever lived. We ignore the suffering affecting our fellow creatures to pursue our own hedonistic definition of well-being. We conspire in ever-more elaborate schemes to control complexity to serve our greed while externalizing any negative impacts. But even after our own death, the punishment continues for those extensions of ourselves—our children—who must continue to bear the consequences of their ancestors' unethical actions.

This comparison to the myth of Sisyphus leads us to an important question: how can we ensure our own ecosystems are heading not towards an eternity that we can't wait to end but to one that we

Evolve or escape— else face extinction.

couldn't imagine how it could possibly get any better? The answer to this lies in our understanding of what suffering actually is. Some believe that suffering arises by the malicious design of an invisible hand. Others believe that it persists only because of a lack of technology. Others chalk it up to the universe or gods working in mysterious ways. I believe that we need to move beyond the search for technology, pain-numbing drugs, or any other scapegoat, and accept what suffering really is:

Suffering is feedback.

FIGURE 5.1. Suffering is feedback for bad (and ugly) design.

Credit: Jarett Sitter

This concept is at least as old as the Greeks, who beyond their allegorical myths, had an explicit proverb *pathemata mathemata* that was equivalent to our modern "no pain, no gain." While you might think this motto is dangerously close to the values of the degenerative and sustainable paradigm, nothing could be further from the truth. Far from being deprecating, shameful, or apathetic, the acceptance of some concept of "what goes around comes around" is the basis of a reverence for all our relations that acknowledges the interconnected well-being of all things. And it is also incredibly empowering! After all, if we are the root cause of our own suffering, then we are the solution. We must refuse to adopt the degenerative paradigm's arrogant faith that humans are exempt from the push-and-pull forces of complexity and evolution. We must reject the sustainable paradigm's belief that humans are somehow irredeemably fallen or flawed. We must stop trying to externalize suffering or escape from its feedback and realize that this planet is our home—there really is no escape. We must evolve, else face extinction. In order to do so, we must monitor our resources for suffering and well-being.

Monitoring Your Resources

Now that I have identified suffering and well-being as our metrics for monitoring our resources, I will introduce some new tools and frameworks. To do this, it is important to remember that our resources are broken into two major categories. The first is our personal resources of financial, material, living, social, experiential, intellectual, spiritual, and cultural capital. The second is the property resources of geography, climate, water, access, structures, fencing, floral, fauna, business, technology, and soil. We'll look at these separately, then pull them together at the end.

Monitoring Your Personal Resources

To start the monitoring process, you need to return to your vision and values one-pager to complete the final component I told you to leave space for back in Step 1, practice 1.3. You now need to identify your vision and values indicators. These are the patterns in your thoughts, emotions, or behaviors that can serve as an early-warning or success system that will let you know if the design and development of your permaculture property is moving you towards suffering or well-being.

> After all, if we are the root cause of our own suffering, then we are the solution.

TAKOTA'S STORY: Monitoring for Mastitis

If necessity is the mother of invention, drudgery is the father of design. This maxim, along with *Drudgery is punishment for stupidity*, have been daily affirmations since I began the design and development of my own permaculture property. And because they never fail to bring about that wonderful laughter of recognition from any audience, I know that other people appreciate them too. Yet, as obvious as these observations of suffering are, the design solutions can sometimes take years to develop.

For example, I began integrating my chicken, pigs, and dairy cows out of a necessity to reduce the amount of time it took to provide for their daily needs and to improve the quality of life of each species. The first goal was met by copying the concept of a work triangle between the fridge, stove, and sink in a well-designed kitchen. My work triangle within the hub of the integrated livestock system consists of feed bins, milking station, and a duckweed pond that represent the fridge (food storage), a water hydrant for the sink (water supply), and five-gallon feed buckets placed under cover outside during the summer and inside the chicken house during the winter as the stove (feed fermentation).

To improve the well-being of my animals, I designed a variety of spokes that radiated from this central hub to create rotational foraging opportunities for all the animals. These spokes currently include a forest garden, cropland, a cattail wetland, an annual garden, and several pastures for the summer rotation. Stockpiled forage, swath grazing, and bale grazing make up the winter rotation. You can see this system illustrated in Figure 3.4 back in Step 3.

This system also makes excellent use of the various animal yields of scratching, shredding, weeding, rooting, trampling, fertilizing, and pest control that further reduce my workload and provide my livestock with meaningful work. It keeps my animals happy and healthy all year-round, and it enables a single person to rotate electric fence to new spokes; milk two cows; feed, water, and prepare the next day's fermented feed for 30 pigs and 150 chickens; and provide bedding in less than two hours per day. You can watch a video case study of this system on the companion website.

Of course, there were lots of little problems, but after many iterations with the 5SP Process, it was shaping up to be not just a good design but an excellent one! That is except for the mastitis.

For the first three years, all of my milk cows were developing frequent bacterial infections in their udders. And each time, it would take several weeks of intensive work hand milking every two hours and massaging their udders with essential oil salves of peppermint and calendula. This might not seem that bad until you realize that mastitis infections cause a great deal of swelling and tenderness, and that even the sweetest of dairy cows can kick very hard. It was drudgery for me and suffering for them.

Taking great pride in the fact we have not used any drugs on our animals for thirty years,

I stuck with our naturopathic approach, and with a great deal of effort, I was able to treat every case of mastitis without the use of antibiotics or any other drugs. Indeed, the treatment was so successful that, despite the drop in production in the portion of their udder that was affected, the next time they freshened, they all milked evenly out of all four quarters—something that rarely happens with the use of antibiotics, as the scar tissue from the infection can petrify their mammary glands indefinitely.

The first year my original milk cow, Sophie, developed mastitis, I just assumed that her immune system was still weakened from her time spent in the industrial dairy farm I purchased her from. However, when her daughter, Blossom, who was born on our farm, developed the same infection, I eliminated this hypothesis. Perhaps they were not getting adequate food, water, minerals, or clean bedding? Experiments ruled all these out as well, and after three years with no sign of improvement, I was starting to get worried. If all my theories about drudgery and suffering were correct, what was the root cause of this problem? Or was this just an unavoidable part of life? I refused to give up, and I kept searching for the stupid thing I was doing.

Finally, after the sixth case of mastitis, a new pattern began to emerge: my dairy cows only developed mastitis when they were in the forest garden spoke. This particular spoke was separated from the hub by a farm road that was used frequently to access equipment and other areas of the farm. If I built a permananent animal alleyway crossing this farm road, it would have created several thousand dollars in expenses and drudgery for me in the form of constantly opening gates to get vehicles across the animal alleyway multiple times a day to access equipment stored on the other side.

In contrast to industrial dairy farms that separate newborn calves from their mothers immediately after birth, I had always kept them together. This provided me with the luxury of never having to bottle-feed a calf and the freedom to leave for several days at a time to teach or consult. But most importantly, it kept the animals happy. From experience, I knew that to separate a calf from its mother causes both animals an undeniable depression. Yet once the calf is six months old, weaning causes no stress to either.

To stop the calf from taking all the milk during that six-month period, I separate the calf from its mother overnight. After the morning milking, I reunite the pair so they spend the day together grazing in one of the spokes, returning in the evening to repeat the process. After a few weeks of training, the calves actually run into their overnight calf shelter and wait for their special treat, a handful of pig feed.

All but the forest garden spoke allowed the cows to return to the hub to visit their calves and make physical contact over the shelter door any time they wished during the night, similar to the way that wild grazing animals cache their babies in a hiding spot away from predators and return to check on them while they snooze. This is why you are never supposed to touch a fawn if you stumble across one out in the woods. It hasn't been abandoned, it has been cached. I was intentionally mimicking this caching in my hub and spoke system. Except for when the cows needed to cross the farm road.

Because I didn't want to spend money on a permanent animal alleyway and then subsequently open gates every time I needed to travel on the farm road, and because herding calves is like herding cats (I had more than a few rodeos to figure this out), I decided the best option was to leave the calves out in the forest garden spoke overnight, lead the cows back to spend the night in the hub, and reunite them in the morning after milking. Even though I knew separating cows from their calves was stressful, I figured this wouldn't be a big deal because they could see each other across the 20-meter (60-feet) wide farm road.

Yet I readily observed both the cows and the calves, while separated in this way, pacing restlessly back and forth, bellowing at each other all night long. I assumed they were bawling because it was a new change and they would get over it. But after three years (I know I'm a slow learner), they didn't seem to be getting over it. They seemed to be getting mastitis.

To test the hypothesis that the change in placement of the calves overnight was causing the mastitis, I finally built a temporary animal alleyway using electric fence so the calves could easily be placed in the overnight shelter and the cows could cross the farm road and visit their calves during the night if they were grazing the forest garden spoke. After dealing with the drudgery of mastitis, opening gates on the animal alleyway didn't seem so bad after all! Actually, to minimize the need to open gates, I subsequently moved the equpiment storage to the north side of the animal alleyway, so there was no choice between the lesser of two evils after all. In fact, all this work took a fraction of the time I was spending treating just one sick cow, so if it worked, I figured it would be a good return of energy invested.

At the time of writing this book, I have had another three years to monitor this new system, and I am pleased to say that drudgery really is punishment for stupidity and the father of design after all! My cows have not had a single case of mastitis since.

Do you remember the five questions introduced in Step 3: Design?
- Is every need of every element met within the system?
- Is every yield of every element used within the system?
- Is every element serving multiple functions within the system?
- Is every function required by the system served by multiple elements within the system?
- Is the system and its elements functioning ethically?

It is clear that some of the emotional or social needs of my cows weren't being met. They were sad and afraid that they couldn't physically touch their calves whenever they wanted. This stress likely weakened their immune system, creating the suffering of mastitis for them and drudgery of treating it for me.

Most farms would have ignored the suffering of the cows and simply used antibiotics to treat the symptoms of mastitis. But this is where the downward spiral starts. Perhaps the cow's microbiome would be disturbed, making digestion difficult or creating opportunities for dangerous microorganisms to get out of hand.

This might contribute to other illnesses that would demand more drugs and drudgery leading to more pollution, and more suffering. All these drugs could be passed through the milk to the calf or humans causing similar effects. The drugs could also pass through the urine and manure to contaminate water and soils. This could lead to antibiotic-resistant superbugs and bioaccumulation of toxins at all levels in the food chain. All this would create ever more drudgery and pollution. All because I didn't want to open a gate.

This experience, and dozens more like it, have galvanized my theory that well-being and suffering are the best indicators of good, bad, or ugly design. While it is almost impossible to create a good design on the first try, by using well-being and suffering as indicators, you cannot fail to recognize and stop an ugly design or to fix a bad one.

For example, some common patterns indicating your well-being is increasing can be feelings of patience, gratitude, reverence, and alacrity. Physical indicators can be falling asleep easily, walking rested and relaxed, and catching yourself unconsciously smiling, laughing, singing, humming, or whistling. Common patterns that indicate suffering can be feelings of anger, resentment, jealousy, and procrastination. Physical indicators can be insomnia, grinding teeth during sleep, or unusually short temper. If you are having trouble observing what your own vision and value indicators are, ask your parents or your spouse. They will know!

One of the clearest indicators of suffering for me is procrastination, or as I call it precognition of drudgery! When it comes to well-being, I experience all of those listed above, as well as something else rather strange. When things are really going well, I am hit with a spontaneous sense of time slowing down, usually accompanied with bursts of gratitude and reverence that bring goosebumps to my entire body. I have heard this sensation described by some as déjà vu. But for me, this isn't a feeling of already having been in a situation I am in, but one of being exactly where I should be.

These indicators of suffering and well-being are like the rumble strips and road signs on a highway. The suffering indicators and the vibrating of your vehicle stop as soon as you get back on track. The well-being indicators and road signs let you know you are going in the right direction. But if you ignore these indicators, you will end up in a wreck or in the wrong place.

The Ten and Two Rule for Monitoring

Continuing this driving analogy further, when I was learning how to drive, one of my instructors shared what she called the ten and two rule. But this wasn't about hand position on the steering wheel. It was about monitoring. Every ten seconds spent looking at the road ahead while driving, she recommended doing a two-second survey of my surroundings that entailed a glance at either the instrument panel or one of the three mirrors. She actually made me count out the seconds! It was certainly awkward at first, but after a few driving lessons, it had become automatic, and I still do it to this day. While some might say spending over fifteen percent of your time monitoring is overkill, I know from experience that this precautionary approach has saved me from a possible life-ending wreck several times.

When it comes to monitoring your personal resources, I have my own ten and two rule, and it takes even less time to achieve the same results. By spending ten minutes, two times a day, plus a few extra minutes at the end of every month (a mere one percent of your waking time), monitoring yourself, imagine how many wrecks or costly detours you could avoid. Imagine how many shortcuts or scenic routes you could discover! To help build a habit around this kind of reflection and monitoring, I recommend two simple things: daily reflections and a monthly vision and values assessment.

Daily Reflections

Back in Step 3: Design, I discussed how deep work is incredibly difficult to perform if you are anxious, depressed, or upset and that making your happiness a priority helps you to better solve problems. Some years ago, Michelle and I were each gifted a simple journal[6] for Christmas by her mother, which introduced the idea of daily journaling. I quickly began to see how beneficial this simple practice was to maintaining gratitude, optimism, and productivity. Inspired by overall improved well-being that we both experienced with these daily reflections, it wasn't long before we were recommending that our clients do the same. The exercise we found to be most helpful consists of writing down the answers to three simple questions in the morning and writing down the answers to three simple questions in the evening.

The morning questions are as follows: What are you most excited about today? What are you most grateful for? What is one thing you could do to make today great? The end-of-day questions are: What is the most amazing thing that happened today? What could you have done better today? What was your biggest insight today? In doing this, you'll find that you are able to:
- help keep your brain in the Goldilocks zone for deep work,
- track insights relating to the design and development of your permaculture property,
- reflect on positive experiences and create a more meaningful following day,
- stay more productive and focused than you would otherwise,
- see patterns of suffering that might otherwise have gone unnoticed, and
- use the exercise as an intuitive scoreboard for the WIG you created in Step 4.

Once you get into the swing of this, you may find, like me, that this little meditation practice is one of the highlights of your day.

Month-end Vision and Values Assessment
The month-end vision and values assessment that I recommend is based on taking ten minutes to review your four or five completed weekly planning pages as well as your daily reflections from the prior month. You then answer four questions:
1. How clear are you about what your weakest link is right now?
2. How confident do you feel that your lead measures and weekly commitments are addressing your weakest link and moving you towards your vision?
3. Have you observed indicators that you and all your relations are moving towards well-being and away from suffering?
4. What has been your most valuable insight in the last month?

Answer the first three questions on a scale of 1 to 10 and compare these to your answers from last month. The final question should be answered with a short journal entry about the biggest insight or patterns between insights that you had written down in your daily reflections. The reason for using a scale and then selecting the biggest insight from last month is that it is the time spent reflecting, not the answer itself here, that is important. It requires more thinking to tell the difference between a 5 and a 6 than a simple yes or no, and it requires a great deal more thinking to read a month's worth of insights and decide which one was the biggest and why. The primary purpose of this assessment is to look for any patterns that indicate either progress or problems in the development of your permaculture property or patterns in your paradigm. By reflecting on the prior month's actions in the weekly planner and daily reflections, you will be able to make better judgments with the clarity of distance. Just like editing your own writing is a lot easier after a month has gone

FIGURE 5.2. The month-end vision and values assessment.

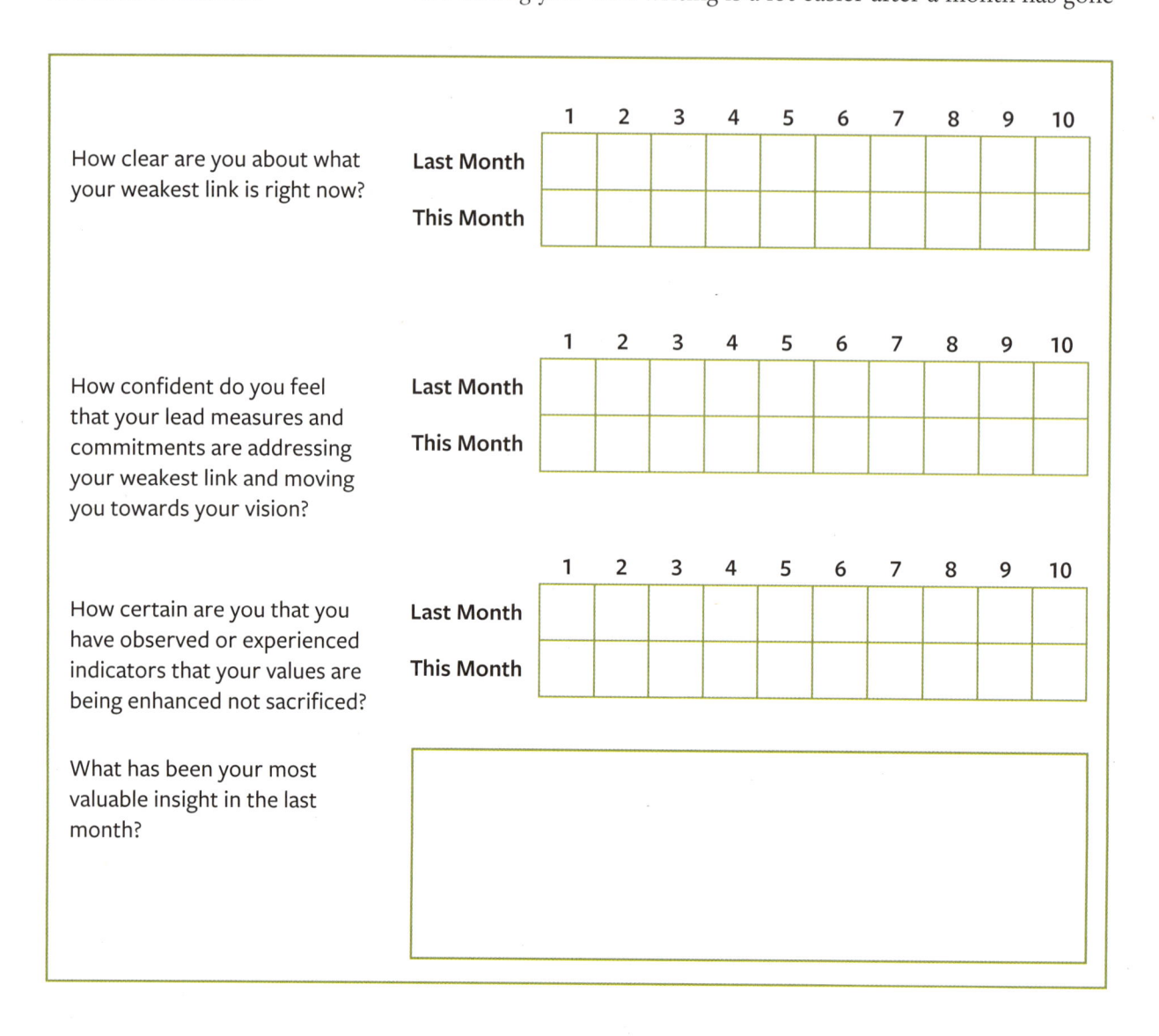

by. You are, in a very real sense, a different person, so you won't take it too personally if you notice some need for improvement, and you are less likely to let it go to your head if things are going great!

Just like counting out loud to build the good habit of checking your surroundings while driving, these monitoring tools will feel awkward when you first start to use them. I promise that, before long, they too will become automatic and help you avoid costly mistakes and take advantage of opportunities. I also realize that there are starting to be a lot of practices, tools, and templates, and you may be concerned about keeping it all straight. If so, don't be distressed as the next chapter will pull everything together and provide some tips for getting and staying organized.

Nature makes a better partner than a slave.

Toby Hemenway,[7]
author and
permaculture teacher

Monitoring Your Property Resources

When it comes to monitoring for the suffering and well-being of your property, I recommend two exercises. The first is doing a yearly re-diagnosis of your property's eleven resources (practice 2.3). In this way, you can compare your annual SWOT analyses to those completed in prior years. Have there been improvements to the weaknesses and threats since last year? How many of your strengths and opportunities have you leveraged? What potential problems or solutions have emerged?

The second exercise is based on ecologist Allan Savory's concept of the **four ecosystems processes**.[8] Shortly I'll provide an exercise that will help you perform this monitoring, but first I need to bring you back to the pinball arcade game introduced in Step 3: Design. Remember the sources, sinks, cycles, shields, and storages? By monitoring the solar energy flow (which is a source), the nutrient cycle (which is a cycle), the water cycle (another cycle), and the successional cycle (a third cycle), you can get a very good idea of the health of your property and the direction it is headed in: upwards towards increased well-being and antifragility or downwards towards increased suffering and fragility.

It can be helpful to think of these four ecosystem processes as the four ingredients of agriculture. Just like a baker uses the ingredients of flour, water, yeast, and salt to make bread, a land steward must use sunlight, soil, water, and organisms to create any kind of a yield. The reason both the baker and the land steward must monitor

TAKOTA'S STORY: Building My Own Permaculture Property

While developing a succession plan for me to take over management of the family farm from my parents, it became apparent that one of the steps needed was for someone to develop a second yard site on the property. After many hours of clarification, diagnosis, and design with my parents, we decided that I would be the one to move out of the second house in the main yard and start anew. This was mainly because my parents really didn't want to go through all the work of building a new property at their age, whereas I relished the opportunity.

This decision was reached in the fall so I had the winter to design before breaking ground with the implementation phase. It was a tight deadline with all of my other commitments, and I knew that I would need my planner and the 5SP Process tools to keep me on track.

My wildly important goal became "to be moved into my new house by next Christmas." I had a few different lead measures to start with, but eventually I settled on "number of hours spent in deep work sessions" where I would lock myself in my office and work with zero distractions for several hours at a time. My weekly commitments usually consisted of getting at least three quotes for whatever aspect of design I was working on at the time. My daily reflections and monthly vision and values assessments demonstrated all the progress I was making and how much I was enjoying every minute of it.

By the time spring came, I couldn't believe how well everything was shaping up. All the pieces were coming together. I had designed every part of my permaculture property, organized financing from my community, and had a detailed project schedule with my critical path activities clearly mapped out. Then everything started to go wrong. An unusually cold winter and late spring meant that the frost was a month late coming out of the ground, setting back the foundation work. The underground power line to the site that had been established twenty years ago shorted out underground and was irrepairable. The only other option of a new grid connection or a solar system put my budget out by more than ten percent and would set the project back another month. All these delays meant that my plans to build a light clay straw house would now be impossible, due to the extremely long drying times of the building materials.

All of these new problems also took time and mental energy to address, and both were in short supply during the busiest season of the farm when animal births and planting gardens, trees, and crops all took precedence. After several weeks of attempting to keep the implementation plan on track, it became increasingly obvious that it was not going to happen. But pride and adrenalin kept me going. It wasn't until a monthly assessment arrived, when I finally had an ah-ha moment.

How clear are you about what your weakest link is right now? Last month it was a 10, this time it was a zero. I no longer felt like material capital, or having a house, was my weak link. Now it was a tie between the social capital and living capital. My relationship with my parents had really been stressed by all of the extra work it was forcing on them, and my body and mind were beyond fatigued.

How confident do you feel that your lead measures and weekly commitments are addressing your weakest link and moving you towards your vision? Last month it was a 10, this time it was a zero. Not only did I no longer have the time to schedule many deep work sessions, but the ones I could fit in went nowhere. No matter how hard I thought, the problems I had encountered were unsolvable, given my existing resources and constraints.

Have you observed indicators that you and all your relations are moving toward well-being and away from suffering? Last month it was a 10, this time it was a zero. My suffering indicators of sleepless nights and grinding teeth now stood out. I also remember experiencing a few anxiety attacks, or at least that's what I think they were, where my heart would race, my head would pound, and I would sweat like I was running a marathon. I have never had one before or since, and I can only assume that this was my intuition trying to get my attention.

What has been your most valuable insight in the last month? One of my insights from my daily reflections that really stood out was: What if I am sacrificing today for tomorrow? This thought had actually been on my mind a lot lately, and it made me nervous. Being caught up in the day-to-day whirlwind, I found it hard to focus on this, but by just taking ten minutes to review my previous four weekly planning pages, daily reflections, and monthly assessment, while reflecting on my vision and values one-pager, it became plain as day. I was sacrificing my vision and values to develop my resources. I was using the wonderful ends to justify horrific means. It took looking back at the past month and comparing it to my vision and values to see this incongruence.

After completing my monthly assessment, I ran the decision to pull the plug through the Good Decision Worksheet and got an emphatic yes. That evening, I brought it up with my parents. I fully expected them to encourage me to try again next year. But it was the exact opposite! They were ecstatic the project wasn't going ahead. They too had been experiencing a great deal of their own suffering indicators because of all the extra work they had to shoulder with me focusing my attention on a new property. Not wanting to change the plan, they didn't say anything and just pushed through as well. Now they were beginning to wish that they could just build a new property that would be easier to care for in their old age. We were all very relieved.

As painful as it was to pull that Band-Aid off, call all my trades, return the loans to my community, and explain to all my friends, family, and colleagues the change in plans, it was the right thing to do. At the time of writing this book, we are still in the design phase of this new retirement property. Regardless if this new plan comes to fruition, I am even more confident now that, whatever happens, as long as I remember to keep using the 5SP Process tools it will be exactly right. Building a permaculture property is inherently random, emergent, self-organizing, and always changing. This complexity can either be unbelievably frustrating or unimaginably useful. I believe the difference lies only in a process and tools. Not to mention the willingness, perhaps even the humility, to actually use them.

their ingredients closely is that they are all essential and interdependent; if any one of them runs out or stops working as well, you're out of business or at least underproducing, no matter how well the others are doing. In other words, your property's well-being, and thus your own, will always be limited by the weakest of these four ecosystem processes. In Step 4: Implement, I described a process for identifying which of your personal resources was weakest so that you could develop a strategy to strengthen it and then move on to the next. The four ecosystem processes assessment described now will help you do the same for your property's resources. Let's first go over each of these ecosystem processes in more detail.

> **Your property's well-being, and thus your own, will always be limited by the weakest of these four ecosystem processes.**

The Solar Energy Flow

Worldwide, photosynthesizing plants draw down hundreds of billions of tonnes of CO_2 from the atmosphere into the soil.[9] This process builds the quantity, fertility, and water-holding capacity of topsoil, which in turn creates the conditions for an increased population and diversity of organisms. However, solar energy flow can be substantially inhibited or, alternatively, substantially increased by your land management practices. Consider Figure 5.3 showing, on the left side of the road, a permaculture farm that emphasizes a diversity of **perennial crops** and, on the right, industrial cropland.

The Permaculture Farm: Land management practices and design decisions for this farm are about enhancing solar energy flow. Rotational grazing, cover/companion cropping, agroforestry, optimized tillage, multi-species covers, and more ensures that the land captures the sun's energy every day of the year on every square foot, unless covered by snow.

The Industrial Farm: The land management practices of monoculture, deforestation, soil erosion, chemical use, and poor water management reduce both the **photosynthetic capacity** (the amount of light intercepted by leaves) and the **photosynthetic rate** (the rate a plant can convert light to sugars). Further, there is an extremely inefficient use of the total solar resource over the course of the year. On May 24, the cropland has only just been seeded to the monoculture of choice for that season. By June 17, which is very close to

[May 24] [June 17] [August 3] [August 29]

the summer solstice (the day of the year with the most sunlight hours available), the crop is just starting to come up, but barely photosynthesizing! In July and August, the plants are finally acting as solar-powered carbon pumps, but by August 29, the farmer has harvested the crop. This field will now sit bare until next year and, during that time, will have zero photosynthetic capacity.

Given how beneficial it is for us to partner with photosynthesis, it's quite shocking to discover that, depending on where you are in the world, cropland has been shown to be bare, on average, anywhere from 30% to 50% of the time.[10]

FIGURE 5.3. Coen Farm (on the left) contrasted with the neighbor's intensive cropland (on the right) in central Alberta, Canada. Photos were taken over the course of several years at different times in the growing season.

Credit: Takota Coen

The Nutrient Cycle and the Water Cycle

The nutrient cycle, which is about taking care of your soil, and the water cycle, which is about taking care of water, are so interdependent that it is much easier to describe them together. Let's start with this very interesting fact: soil carbon can, on average, hold four times its weight in water.[11] If you think back to the story about Takota's dam and swale (in *The Good, the Bad, and the Ugly*), in the section Wrong Scale), this is why he did not get the design quite right on the first go-around.

This one-to-four ratio means that soil water-holding capacity increases dramatically with even a slight increase in soil carbon content. For instance, a 1 percent increase in soil carbon on your property will store an additional 17 liters of water per square meter (0.4 gallons per square foot).[12] But the opposite is also true: the same amount of water-holding capacity is lost when carbon is lost. In large part, it's our soil and nutrient cycling practices that are directly responsible for either carbon gains or carbons losses, directly correlating to water gains or water losses. Note that agricultural soils around the world have lost between 30 and 75 percent of their carbon, adding approximately 80 billion tons of carbon dioxide to the atmosphere in the last 150 years.[13] We've basically set ourselves up for these carbon-degraded lands to require more irrigation than they'd need otherwise, while increasing the severity of drought as well as their vulnerability to erosion. Now think back to Figure 5.3 and this time consider the differences of the nutrient and water cycles.

> The **Permaculture Farm** emphasizes land management practices that maintain and support life, and carbon, in the soil. Actions that may inhibit or harm soil biological activity are minimized, or avoided, such as tillage. Water runoff is very low, and any runoff is clear or very slow and spread out. Wetlands are preserved, new springs are forming, and aquifers are being recharged.

> When it rains, the bare (fallow) and compacted fields of the **Industrial Farm** erode into a deeply incised stream. The stream water leaving the property is dirty and contains residues from fertilizer, fungicides, insecticides, and other chemicals. Wetlands have been drained to increase the planting area for the monoculture crop.

Perhaps you've noticed that many of the practices compared for nutrient and water cycle are, in fact, repeated from the previous section on maximizing the solar energy flow. This illustrates the real-world connectedness and interdependence of these cycles, as well as the momentum that can be built and the upward spiral that can be initiated as we design with these cycles in mind.

There's definitely far more to both the nutrient and the water cycle than just soil carbon, and I encourage you to do your research and learn more. Soon I'll share how to monitor your property to ensure that these cycles are indeed upward spirals. For now, let's move on to the fourth ecosystem process, the successional cycle.

The Successional Cycle

A particular pattern occurs when the solar energy flow, the nutrient cycle, and the water cycle interact. This is known as the succession cycle. All ecosystems are born from a disturbance like a fire, landslide, volcanic eruption, flood, or an exploded star. After this death, pioneer organisms begin to arrive or evolve, and over hundreds, sometimes thousands, of years of research and development, these flora, fauna, fungi, and other organisms of a particular place arrive at a climax ecosystem that is diverse, stable, and highly productive. In short, this climax represents the optimal expression of the other three ecosystem processes based on a particular geographical and climatic context. It is what your land wants to become if left alone to its own devices. This climax might be a forest, a savannah, a prairie, or a desert, but it is the best result after billions of years of trial and error. It is for this reason that, wherever you are in the world, your climax ecosystem is your blueprint for success and the benchmark you should compare your own property against. Because if your property is doing better without you, you must be doing something wrong!

I must stress that the successional cycle is just that, a cycle! Perhaps it is the fear of our own death that has led to the phobia of changes in ecosystems, but whatever the case, it is futile to take a snapshot of an ecosystem at a specific time and then use force to hold it to that image of its former self, labeling any evolutions as pests, or invasives. Ecosystems, just like everything else in the universe, have their own births and deaths. In other words, the circle of life is not a straight line towards a better and better climax; it is an upwards spiral, that sometimes, to our limited perspective, can look

Your climax ecosystem is your blueprint for success and the benchmark you should compare your own property against.

like it is going backwards. The goal of permaculture is to partner with this successional cycle and, through the use of optimized disturbance events, speed up its natural progression towards increasing complexity and well-being.

Let's bring it back one more time to Figure 5.3 and compare how these two farms manage the successional cycle.

The **Permaculture Farm** absolutely embraces biodiversity. An increasing number and diversity of organisms is seen as success. Livestock and animals are viewed as an integral part of a healthy system through practices such as nutrient cycling, pasture cropping, and planned grazing.

FIGURE 5.4. The successional cycle (also known as ecological succession) is a pattern that occurs when the biological community evolves through time.

PIONEER ⟶ CLIMAX

Credit: Jarett Sitter

The **Industrial Farm** considers other living things as pests or as threats to productivity, where productivity is measured in bushels per acre of one single crop for the year. Large swaths of trees and shelterbelts have been removed, and with no habitat, very little wildlife is found onsite.

Monitoring the Four Ecosystem Processes

Now that I've briefly explained the four ecosystem processes, know that the goal when monitoring your property's resources is to look for signs that indicate you are optimizing solar energy flow and creating an upward spiral through your management of your nutrient, water, and successional cycles. Figure 5.5 shows the template I recommend you use, and just like your personal vision and values indicators, these indicators present a clear sign that the well-being of your land and thus you is increasing.

Complete this template once every three months, or once per season, for several locations on your property. At a minimum, choose a location that you are actively managing as well as a **zone five** area on your property. This is why every property needs a wilderness area or a zone five that is unmanaged by humans. Without it we have no classroom to learn from or an answer key to check our work against. If an ecosystem without humans is doing a better job managing the four ecosystem processes, then you have some evolving to do! If your own zone five isn't a climax ecosystem, then you'll have to seek one out so that you can thoughtfully reflect on both your blueprint and your benchmark.

The practice is simple, and you need only to review the indicator and give a qualitative rating on a score of one to ten. Ten, the highest score, represents "as good as my climax ecosystem," and one is the lowest score, meaning that you have a lot of work to do! Take note that every single one of these indicators does not require anything more than your own eyes and hands to observe and a pen and paper to record. At most you will need a shovel if you are unable to dig in with your hands. We have done this intentionally to make monitoring easy.

> **The goal when monitoring your property's resources is to look for signs that indicate you are optimizing solar energy flow and creating an upward spiral through your management of your nutrient, water, and successional cycles.**

FIGURE 5.5. Use this template to monitor soil, water, organisms, and sunlight on your property. For each statement, provide a rating on a scale of 1 to 10, where 1 is "not at all" and 10 is "as good as my climax ecosystem." Adapted from Allan Savory and his concept of the Four Ecosystem Processes.[14]

Nutrient cycle	Rating
Good mulch layer covering the soil.	
Soil structure is porous, friable, noncompacted, and high in organic matter.	
Soil aggregates are large and round, not blocky or angular.	
The soil is always moist, but never waterlogged.	
Fast decomposition rate of manure and plant material, no oxidation and lignification is present.	
Mycelium hyphae are visible.	
Soil coating is visible on root hairs.	
Many macro soil organisms are visible.	
Soil erosion is far less than soil creation.	
Minerals are cycling from a full soil profile via both deep- and shallow-rooted plants.	

Successional cycle	Rating
Diversity of organisms is high, stable, and increasing.	
Organisms are healthy and disease free.	
Species of fauna that are larger in size and with longer life cycles are becoming more common.	
Species of flora that are larger in sizes, with less aggressive characteristics and longer life cycles are becoming more common.	
Top ecosystem predators are present and becoming more common.	

Water cycle	Rating
Precipitation is slowed, spread, and infiltrated rapidly.	
Water runoff is very low, and what movement there is, is very slow and spread out.	
Any water that runs off the property is clear.	
Evaporation and sublimation losses are minimized.	
Condensation (dew) is prevalent in the mornings and evenings.	
Both droughts and floods are becoming less frequent and less severe.	
Springs, streams, aquifers, and wetlands reappear and have steadier flow rates, increased supplies or increased residence times.	

Solar energy flow	Rating
Soil completely covered by living plants.	
Any soil not covered by living plants is covered by dead plants, a.k.a. mulch.	
Vigorous and varied root systems on plants.	
Vegetation growth rates are very rapid.	
Photosynthesis occurs over as long a time period as possible.	
Overall high diversity of flora and fauna (think in scale, timing, placement, and form).	

TAKOTA'S STORY: An Ecosystem Disguised as a Farm

Imagine planting your usual vegetable garden on the May long weekend. Now imagine taking a holiday for a month. What would it look like when you got back? How about after a year? A decade? A century?

The time-lapsed images of "ecological chaos" fluttering through your mind represent the evolution of your ecosystems through time. At some point, this biological community would evolve into your climax ecosystem or biome. It is the highest expression of natural succession that can be achieved, based upon various factors like your property's eleven resource categories.

Here in central Alberta, our 100-hectare (250-acre) farm resides in the aspen parkland biome, which consists of groves of poplar, spruce, shrubs, vines, and ground cover species interspersed with grasslands that were historically home to vast herds of migratory grazing animals. In fact, there were an estimated 30 to 60 million bison alone in North America prior to European settlement, and no one fed, sheltered, watered, or checked them at 3 am during calving season.

So, if this is what humans can do without all the gadgets of precision agriculture,[15] why would we do anything else? It was this question that led me to pursue extensive research into forest gardening, permaculture design, and anything to do with biomimicry. The short answer to my search is that it is entirely possible, even easy, to identify patterns from Mother Nature's 4.5 billion years of research and development to develop appropriate agricultural systems that embrace and even enhance the tendency for ecosystems to evolve effortlessly based upon their specific context. The best way to do this is to mimic your successional cycle.

So what exactly does that look like? Using our aspen parkland biome as a blueprint for success on our farm, we can swap poplar, birch, and box elder for apples, pears, and plums. Change our native conifers for Korean pine, and we have pine nuts as big as pistachios! Throw in some walnuts, butternuts, and oaks, and we have our canopy species. The mid-story and understory species are a little easier as hazelnuts, raspberries, cherries, saskatoons, cranberries, gooseberries, and currants are all delicious natives already. However, we could add a few extras like honeyberries and hardy kiwis just to mix things up a bit. For the ground layer, we can stick to strawberries and an incredible variety of medicinal plants that already call this place home.

That covers our wooded areas, now what about the interspersed areas of grasslands? Simply plant native grasses and legumes in between our groves of fruit and nut trees, and we are almost done. The last step is integrating animals into our farm system. If we pull our cattle, hogs, and poultry out of our barns and CAFOs, putting them back on the land where they belong, they can manage the grassland and heal our ailing nutrient cycle. Add in all the native birds, amphibians, reptiles, insects, and mammals that are desperately looking for home sweet home, and you have it: a thriving ecosystem cleverly disguised as a farm.

Yeah, but what do they actually produce? These farm ecosystems can provide all the necessary carbohydrates, proteins, and oils for human and animal consumption, fruits and berries for vitamins and minerals, plants for medicine, trees for timber, fuel, and forage, along with a myriad other ecosystem services. We currently grow 90% of our own calories and provide 50+ local families with the majority of their fat and protein needs. In fact, based on my calculations, if all of the farmland in Alberta followed our lead, we could feed the population of our province many times over.

Ok, great idea, but does it pay the bills? Five years ago, I went to my community with a plan to develop a farm such as this. I called it a Forest Garden Community Supported Agriculture (CSA). Unlike in a typical vegetable CSA that only lasts a season, our community members purchased $500 "investments" that were paid back in farm products over the course of four years with a 110% return of value. With the $10,000 generated, I was able to purchase over 5,500 fruit, nut, and berry perennials. The investors even came out to help plant.

By the fifth season, the yield from our 2.5-hectare (6-acre) forest garden had already paid off the original financial capital investment. Not bad, considering that it will continue to increase in yield and diversity for many decades with only minimal maintenance. This forest garden, along with our direct-marketed milk-fed pork, grass-fed beef, and pasture-raised eggs, currently provides a living wage for my parents and myself, despite the fact we are still rebuilding our new yard site and that most of our enterprises are still in their infancy in terms of production. The best part is we have never worked less!

FIGURE 5.6. An ecosystem disguised as a farm. Photo taken in fall 2020.

Our current agricultural system is fighting the natural evolution of ecosystems with biocides and tillage, expending huge amounts of time and energy in the process. This overly simplistic farm design has led to the degradation of

Credit: Takota Coen

our air, soil, water, and biological diversity and the proliferation of drudgery. We are literally eating our planet to death, all while going broke and working ourselves to the bone. This doesn't have to be the case! By paying attention to our local biome, we can design our farms in Nature's image for nutrient-dense food, healthier land, stronger communities, financial freedom, and the time to actually enjoy life.

Practices for Step 5: Monitor

Here are the practices for Step 5: Monitor. You'll notice that the practices related to monitoring all recur on a recommended time-step, and as such, a planner (discussed in the next chapter) is an essential tool. As always, templates and downloads are available at mypermacultureproperty.com.

Practice 5.1: Create Your Vision and Values Indicators

Take out your vision and values one-pager, from practice 1.3, and at the bottom of the page, write down a few positive and negative indicators (patterns of thoughts, emotions, or behaviors). Here are a few common examples:

Positive Indicators:
- I sleep easy and wake rested.
- I feel balanced and happy.
- I feel passionate and cheerful toward my work.

Negative Indicators:
- Sleep is difficult, and I'm waking tense and sore.

Monitoring Tools for Google Earth Pro

In Step 2, I recommended that you get a GPS app for your mobile device or smartphone, specifically one that allows for output in a format suitable for Google Earth Pro (GEP). This combination of tools and software (GEP + mobile device + GPS app) also supercharges your monitoring ability. When performing your quarterly monitoring, or whenever you see something worth recording, you simply snap a photo using software that allows you to include a geographic position and that allows for easy export into what's called a KML or KMZ file, the format needed for GEP. GPS Kit or GPS Essentials are two of many different smartphone apps that can do this. You can also record georeferenced text-based descriptions. When you import the photo (or text, or coordinates, or routes, or any other pertinent geographic information) into GEP, it will appear as a layer at the correct location on your property map. And if you made sure that the date and timestamp settings are also enabled, this visual information can become part of a cumulative stack of images that help you to better see trends and patterns at specific locations, over long periods of time.

- I feel anger, resentment, and/or jealousy.
- I feel procrastination and drudgery in my work.

Practice 5.2: Reflect and Record

Start your day by writing down answers to the following questions:
- What are you most excited about today?
- What are you most grateful for?
- What is one thing you could do to make today great?

End your day by writing down answers to the following questions:
- What is the most amazing thing that happened today?
- What could you have done better today?
- What was your biggest insight today?

Practice 5.3: Month-end Vision and Values Assessment

Every month, take five minutes to answer the following questions on a scale of 1 to 10. Compare your answers to the prior month.
1. How clear are you about what your weakest link is right now?
2. How confident do you feel that your lead measures and weekly commitments are addressing your weakest link and moving you towards your vision?
3. Have you observed indicators that you and all your relations are moving towards well-being and away from suffering?

Also, take another five minutes to review your daily reflections from the past month (practice 5.2) and record your most valuable insight for the month.

Practice 5.4: Monitor the four ecosystem processes

Build yourself a checklist for monitoring each of the four ecosystem processes by copying Figure 5.5, or by downloading a version from our website. Complete this template once every three months (or once per season) for several locations on your property. At a minimum, choose a location that you are actively managing, as well as a zone five area. I also recommend that, at least once per year, you visit a healthy climax ecosystem to leverage as your blueprint and your benchmark.

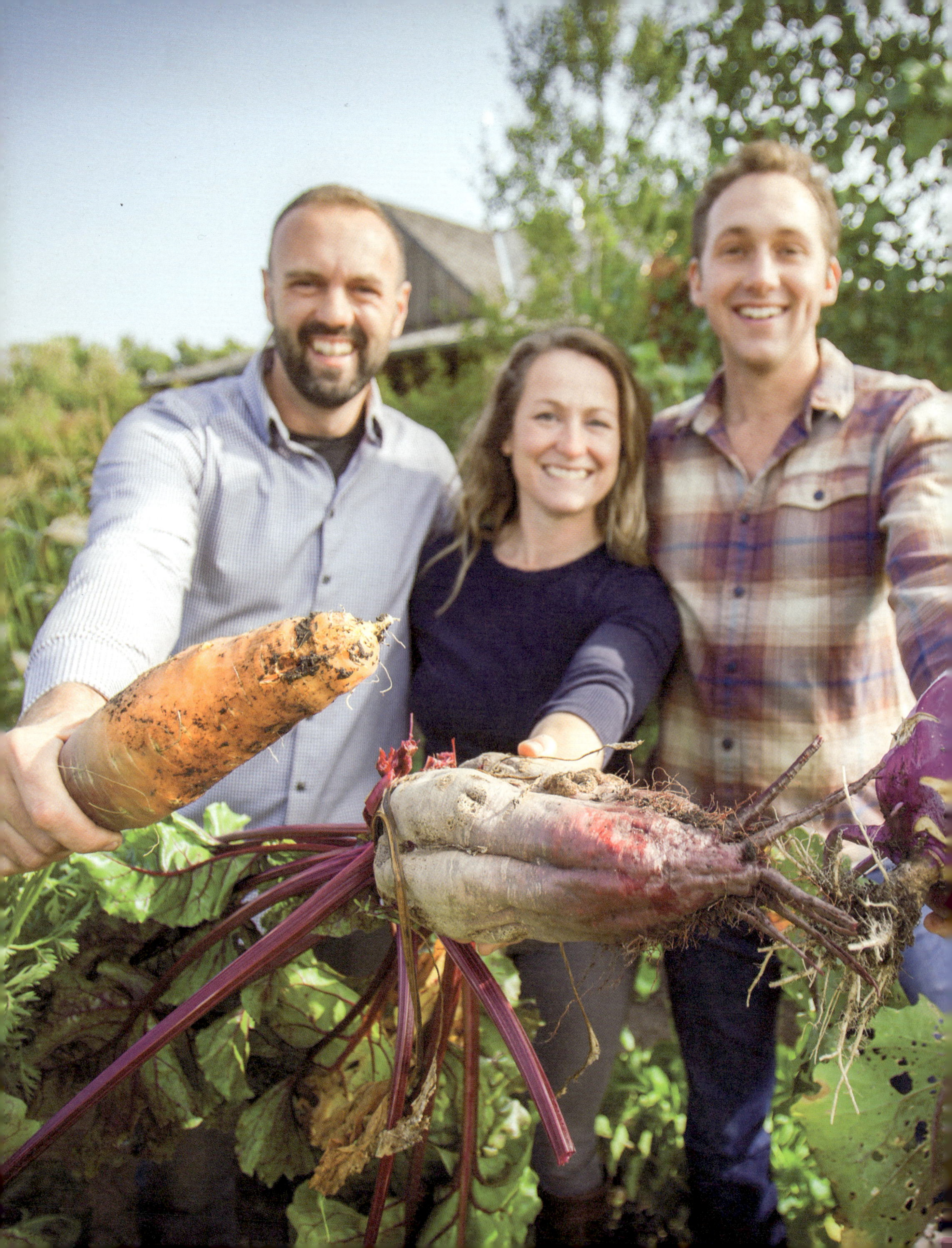

The Solution to a Sisyphean Task

When you really start to monitor your personal and property resources with full honesty, you will likely find, as I have, that there is a great deal more suffering than well-being in our world at present. This is a soul-crushing realization, but I see it as a necessary one because it leads to earnest action.

Credit: Kurt Stenberg

> *We must find our way back to true nature. We must set ourselves to the task of revitalizing the earth. Regreening the earth, sowing seeds in the desert—that is the path society must follow.*
>
> Masanobu Fukuoka,[1] Japanese farmer and philosopher

So, what exactly do you do if you find yourself already having been punished to an eternity of pushing your rock of a permaculture property up a mountain of your own making? What do you do if you find yourself in a world that is filled with unbearable and seemingly unlimited suffering?

Our only hope for our own present Sisyphean dilemma lies in an ingenious solution proposed by a German philosopher.[2] Instead returning downslope empty-handed to retrieve his boulder, Sisyphus should break off a hunk of the mountain, carry it down to the lowest point, and exchange it for his runaway rock. Given the unlimited time span, it would only be a matter of time before the mountain and the valley would be level and Zeus's enchanted rock would have nowhere else to roll. Perhaps his efforts would please the god of thunder and lightning, and he would grant Sisyphus the ability to spend the rest of eternity as he wished. Although, I'm sure this would come with conditions, and under the eye of a very watchful host of gods.

The only way to overcome the many mountains of suffering on our own permaculture properties and across the globe is by using the same persistence and process. We must start to dismantle the mountain piece by piece.

In the same way you don't have to know everything before you take the first step on your own permaculture property, we, as a society, don't have to sort out all the details of how we are going to fix all the nuclear waste, landfills, salinated soil, chronic disease, or government corruption. The process to figure out the solution is the same for both: clarify your vision, values, and resources; diagnose your resources for strengths, weaknesses, opportunities, and threats; design your resources to meet your vision and values; implement the best design that will most improve your weakest resource; and monitor your resources for the indicators of well-being or suffering.

> *There is nothing in a caterpillar that tells you it's going to be a butterfly.*
>
> R. Buckminster Fuller,[3] architect, systems-thinker, and author

This may take many generations, but just like Sisyphus, we have a time span close to eternity at our disposal. We can either accept endless suffering with zero chance of well-being, or turn our Sisyphean task into a strategic labor of love that turns the problem of suffering into immediate and unimaginable well-being.

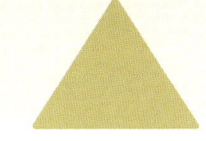

Putting It All Together

I've now presented all of the steps that will help you to overcome the struggles of *What to do*, *Where to find*, *How it all connects*, *Where to start*, and *When will it end?* It's time to both recap and put it all together. I've also shared practices and templates for each step, and some of these practices have timelines and should be repeated daily, weekly, monthly, quarterly, or annually. You may already be thinking, *how am I going to keep track and execute all of this?*

The solution is simple. You need a system to plan, prioritize your tasks, track timelines, and keep you organized. You need a good old-fashioned paper planner. Despite a recent trend to move to digital

FIGURE 6.1. The only way to overcome the many mountains of suffering on our own permaculture properties and across the globe is by using persistence and process. We must start to dismantle the mountain piece by piece.

Credit: Jarett Sitter

tools, loads of research[4] shows that a physical planner is far better at helping you stay organized, develop good habits, have new insights, and notice trends, as well as get your brain into focus mode.

Many off-the-shelf planners are available at your local stationery store and online. Look for one that provides enough space to record the daily responses to the six questions (practice 5.2) and the weekly responses to the activities from Step 4, including your weakest link analysis, wildly important goal, lead measures, and weekly commitment (practice 4.1). Once you have chosen a planner, go through the whole calendar year and note the dates when you should be completing the other practices, as summarized in Table 6.1. Glue or staple a copy of your vision and values one-pager into the inside cover so that it's handy. If the planner has a pocket, you may also

You need a system to plan, prioritize your tasks, track timelines, and keep you organized.

FIGURE 6.2. The most fundamental tool in the 5SP Process is a good old-fashioned paper planner.

FIGURE 6.3. The real-life evolution of my, Takota's, and Michelle's planning tools.

want to slide in a few copies of the Good Decision Worksheet (practice 4.2) so that it is easily available if you need to make a decision.

Instead of purchasing a planner, you can also print multiple copies of the templates we provide and put together your own booklet, whether you staple it or get it bound at the local stationery. In fact, both Takota and I did exactly this for many years as we couldn't find a style of planner that was just right. Our weekly planning page (template 4.1) is actually based on a layout that Michelle built for herself over fifteen years ago because she couldn't find an off-the-shelf version that worked for her, and she has used this template religiously since then.

As a third alternative, we have built an official *Building Your Permaculture Property Planner* as a companion to this book. It is designed to help you with the exact daily, weekly, monthly, and quarterly practices we recommend in the 5SP Process, in addition to providing space for your own planning and to-do lists. Learn more about this on our website, mypermacultureproperty.com.

I have found from my own experience and work with clients and students that it takes using a planner and all of these tools for at least a full year before you can truly see the benefits. After that, you may find that you have developed the monitoring habits that make parts or even all of the planner redundant. But don't make that call until the year is out. By that time, your own suffering and well being indicators will let you know if it's worth continuing to use!

The life I am currently living is a thousand times more closely aligned with what I really want and what is important to me than it was merely a decade ago. I attribute this significant improvement in alignment to five factors. The first is that Michelle and I invested the time early on to get clear on our vision and values, and we put down, on paper, *how we really wanted to live,* given our limited numbers of hours on this planet. The second is that we took inventory of the resources we had access to, for our unique situation, in order to achieve our goals. Then, third, we learned about permaculture and applied this methodology to start creating a design that was right for us. The fourth is that we consistently used an execution and decision-making framework to ensure that we were focused in the most effective place and that our big decisions would continue to propel us towards our vision. Finally, we used a day-to-day planner to make sure that we stayed productive, optimistic, focused, and on

> I have found from my own experience and work with clients and students that it takes using a planner and all of these tools for at least a full year before you can truly see the benefits.

track. Quite simply, these five factors make up the backbone and the steps of the 5SP Process: Clarify, Diagnose, Design, Implement, and Monitor.

Just over a decade ago, as an oil and gas engineer, I could have never imagined where I would be now, a full-time permaculture consultant and educator currently building my own farm-scale permaculture property. I have come to respect the power of having a vision and the immense value of the tools that keep me on track as I continue to reach for this vision. Having invested thousands of hours into distilling what has worked for me and my clients, my advice to you is this: clarify your vision, values, and resources and diagnose your resources for strengths, weaknesses, opportunities, and threats. Design your resources to meet your vision and values. Implement the right design that will most improve your weakest resource and monitor your resources for indicators of well-being or suffering. Be doggedly persistent through all of this, follow the process, and as my mother-in-law always says, "You'll overestimate what you do in one year and underestimate what you can accomplish in five."

> **You'll overestimate what you do in one year and underestimate what you can accomplish in five.**

Your Very Last Practice: Your Permaculture Property Planner

The very best way for you to stay productive, optimistic, focused, and on track is to put everything I've shared with you together in your own calendar and planner booklet, which you can acquire by one of the following options.

- Head to the calendar and planner section of your local stationery store or simply do an internet search for "calendars and life planners." Look for one that provides enough space to plan your day, answer the daily questions from practice 5.2, and fill in the weekly activities/reflections from practice 4.1. Go through the entirety of the calendar year and note the dates when you should be completing the practices, as summarized in Table 6.1. Glue or staple a copy of your vision and values one-pager into the inside cover and keep a few copies of the Good Decision Worksheet (template 4.2) in the pocket or folder under the back cover.

TABLE 6.1. The Five-Step Permaculture Process. Summary of the practices, templates, timing, and other recommended tools for the 5SP Process.

Practices	5SP Process Templates/Tools	Timing/Interval	Third-party Tools
BEFORE GETTING STARTED			
Get an Accountability Partner		At the outset	
Step 0. CHECK			
Practice 0.1: Reflect on major influences, biases, and assumptions you may hold		At the outset	
Practice 0.2: Study, observe, and contemplate ecology		At the outset	
STEP 1. CLARIFY			
Practice 1.1: Reflect on what you have, what you want, and what is right		At the outset	
Practice 1.2: Create (and maintain) your personal resource inventory	Template 1.2: Resource Inventory	At the outset. Review (and update) when making major decisions	
Practice 1.3: Create your vision and values one-pager	Template 1.3: Vision and Values One-Pager, Template 1.3b: Core Values Word List	At the outset Review it daily for at least two months	
STEP 2. DIAGNOSE			
Practice 2.1: Organize your existing information	Template 2.1: Computer Folder Filestream, 5SP Process Filestream for Google Earth Pro	At the outset	
Practice 2.2: Gather new information		At the outset, and ongoing	Mobile device (smartphone) with GPS app or handheld GPS device
Practice 2.3: Complete a SWOT on your personal and property resources	Template 2.3a: Personal Resources SWOT Diagnosis, Template 2.3b: Property Resources SWOT Diagnosis	At the outset Review and re-diagnose once per year	

TABLE 6.1. (cont'd.) The Five-Step Permaculture Process. Summary of the practices, templates, timing, and other recommended tools for the 5SP Process.

Practices	5SP Process Templates/Tools	Timing/Interval	Third-party Tools
STEP 3. DESIGN			
Practice 3.1: Set yourself up for deep work		At the outset, and ongoing	
Practice 3.2: Study case studies		At the outset, and ongoing	
Practice 3.3: Create your adaptable and interactive permaculture property design		At the outset, and ongoing	Google Earth Pro or other GIS program
Practice 3.4: Consider the five testing questions and design for energy flows	Template 3.4a: Five Questions, Template 3.4b: Design for Energy Flows	At the outset, and as required/useful	
STEP 4. IMPLEMENT			
Practice 4.1: Plan your week, weakest link analysis, set a wildly important goal, describe lead measures, and make a commitment	Template 4.1: Weekly Planning Page	Monday mornings or once per week	
Practice 4.2: Make good decisions, and avoid bad and ugly ones	Template 4.2: The Good Decision Worksheet	As required, when making decisions	
Practice 4.3: What are your birdshots?		As needed to spur ideas	
STEP 5. MONITOR			
Practice 5.1: Create your vision and values indicators		At the outset	
Practice 5.2: Reflect and record		Morning and evening every day	
Practice 5.3: Month-end vision and values assessment	Template 5.3: Month-end Vision and Values Assessment	End of every month	
Practice 5.4: Monitor the four ecosystem processes	Template 5.4: Monitor the Four Ecosystem Processes	Winter solstice, spring equinox, summer solstice, fall equinox (or once per season)	Mobile device (smartphone) with GPS app or handheld GPS device

- Print 52 copies of template 4.1, 12 copies of template 5.3, 4 copies of template 5.4, one copy of your vision and values one-pager (template 1.3), one copy of each SWOT analysis (template 2.3a and 2.3b), and 12 copies of the Good Decision Worksheet (template 4.2). Staple together this booklet or get it bound at the local stationary. Note that you can use the blank/back side of the weekly planning page (template 4.1) to write down answers to the daily questions (practice 5.2). This method also allows you to customize your to-do lists and daily timetable in whatever layout works best for you.
- As a companion to this book, the *Building Your Permaculture Property Planner* is designed to help you with the exact daily, weekly, monthly, and quarterly practices we recommend here, in addition to providing space for your own planning and to-do lists. Learn more about this on our website, mypermacultureproperty.com.

FIGURE 6.4. Our best advice for buiding your permaculture property is to follow this process: clarify your vision, values, and resources; diagnose your resources for strengths, weaknesses, opportunities, and threats; design your resources to meet your vision and values; implement the right design

that most improves your weakest resource; and monitor your resources for indicators of well-being or suffering. If you do this, you may not know where you'll end up, but we guarantee you'll want to be there.

Credit: Jarett Sitter

Afterword:
The Land Needs Us
to Live Differently Here

This book is our attempt to answer a single question: How can we, as individuals and as a species, live a good life without cutting our branch in the tree of life out from beneath ourselves?

A profoundly simple answer to this question was given to Takota in a chance meeting with an Indigenous Elder of the Dené people named René at a conference in 2016. After seeing Takota present about his attempts to answer this question on Coen Farm, he approached him during a coffee break and expressed an interest in sharing a story about how the Peace River (a large river originating in northern British Columbia) got its name. Takota eagerly accepted his gracious offer. René's story began with a summary of the displacement of the Cree people from the Great Plains of southern Alberta during the 1800s. European settlers and various land treaties had already begun to push the Cree further north into lands covered in thick boreal forest. However, the Cree were met with resistance from the Dené people who had called these forested lands home for centuries. Fierce battles were fought. Many warriors died on both sides. When both groups recognized that an easy victory was not possible before the coming winter, Elders from the two warring tribes agreed to meet on the banks of a river to negotiate terms of a peace treaty so that all could begin preparation for the waist-deep snow and minus 40° cold. It was decided that the river upon whose banks they had gathered would now serve as the dividing line of each tribe's new territory. The Dené people would stick to the northern side of the river, while the Cree were given the lands to the south. Hence the name Peace River. An interesting piece of trivia Takota thought, but René's story didn't stop there.

That winter, the Cree people were on the brink of starvation. Each hunting party that departed their large camp into the dense boreal landscape had to travel farther and farther, and each time, they came back with less and less of the moose, elk, and deer that their people needed to survive. When the hunters came back with nothing, the entire camp would relocate elsewhere in the forest, and the process of diminishing returns would repeat itself. As the winter progressed, the deep snow and fierce cold began to hinder these relocations. Finally, it was decided that one of the Elders would embark on a vision quest to seek guidance from the spirits about what their people should do to live in this new and unfamiliar land.

After four days, the Elder returned, and the entire camp gathered to hear of his vision. He began with the proclamation, "The land needs us to live differently here." He then continued by sharing the details of his vision. In his vision, he was shown that his people's methods of hunting herds of bison, developed over thousands of years on the open plains, would not work in the dense boreal forest. Their large and mobile population that was once essential to track their prey, construct funnels along jumps or impoundments, and efficiently process the enormous quantities of meat and hides that would come in short bursts following a successful hunt were now a warning beacon to their solitary prey in the forest. The elder was shown that, to live in harmony in this new place, the land needed them to learn a new way of life. The Elder finished by saying, "The land needs us to become like our Dené brothers and sisters" and that his people would need to disband their single large camp and begin to live in many small villages distributed at great distances away from one another so that the animals would no longer fear them.

Upon hearing the Elder's vision, the entire camp broke into murmurs. They all knew in their hearts that this was their only way to survive in this new land, but they also knew it meant something else. All of their skills, and most importantly their close-knit relationships to each other, had, for eons, developed around a single large camp where their neighbor's tepee was only 100 paces away, rather than 100,000. To live differently meant the death of their very way of life. The Cree were inconsolable. Suddenly, a messenger from a nearby Dené camp across the river appeared in their midst. The entire camp of the Cree people surrounded him and began to question the messenger about his people's way of life. How did they hunt?

What kinds of tools and shelters did they use? And most importantly, how did they manage to cope in such a desolate place so far apart from their kin? Happily, the Dené messenger began to share his people's way of life. He also explained their elaborate system of messenger "runners" that facilitated the sharing of resources and information throughout the year and how large seasonal gatherings at central locations maintained the kinship of their entire population. The Dené messenger assured his Cree brothers and sisters that this was a wonderful land and that his people would share their way of life so that they too could thrive here.

René's story was cut short by the start of the next session, and Takota never saw him again. After trying to connect for over a year, we learned through his sister that he had passed away. We never learned how the pioneering Cree fared during their first winter in their new land or the deeper message René was trying to teach Takota that day.

In the ensuing years, we have all thought much about this story, and we believe that René was trying to teach at least two things that day. The first being that humans have always struggled to answer the question of optimal and enduring well-being in a complex world regardless of the color of their skin, the gods they pray to, or the level of their technology. The second being the similarities mirrored in our present situation. As then, we are living in a strange new land. There are clear indicators that our people are outstripping the resource base upon which we depend. Our own hunters are spending more energy in search of resources than the resources themselves are worth. Our winter is approaching. Our Elders have been sharing the same vision of hope. And many messengers have arrived in our camp to share much-needed knowledge about how to live differently.

If this was René's intention, then we couldn't agree more. However, in writing this book, we have come to wonder if there wasn't something else the Cree in that camp in the 1800s shared with us today. We wonder if, after hearing the vision of their Elder and the knowledge from their messenger, they also shared our paralyzing fear, ignorance, and naive optimism regarding the state of our own resources, the promise of new technology. Perhaps some of the Cree shared the same temptation to stay put in their old familiar way of life. Perhaps some of them had tried to live differently in this new land but, failing, had returned to their old camp. If this was the case, then we are sure that, as their brothers and sisters had gained new knowledge, they would have sent messengers back to the old camp to share their firsthand experience and compel their kin to join them.

We have listened intently to the many Elders, teachers, and messengers already living well in this new land. We have had our struggles, but are now living well here. We know we are not the first messengers to return to our kin, nor will we be the last. Our greatest hope is that our message, this book, will join the chorus of all those that have come before us and provide additional aid in our shared and ongoing attempt to learn to live differently on the land.

— Rob Avis, Michelle Avis, & Takota Coen

Glossary

Alluvial fan: Geomorphic feature that is located at the depositional end of an erosional-depositional system, generally at the bottom of the sloping surface that connects a mountain to the surrounding plains. Their form is a function of erosion and subsequent transfer of rock debris from one portion of a watershed to another. Deposits tend to be fan-shaped from above and are best described as a segment of a cone radiating from one single point source.

Complex system: A system containing interconnected and interdependent elements organized in a way to achieve some overarching function or purpose that also has the capacity to self-organize and evolve.

Downward spiral: A series of thoughts/actions that feeds into itself, causing a situation to become progressively worse.

Energy slave: The equivalent energy provided by technological mechanisms, modern machines, and energy, in terms of the amount of work one fictive human could perform.

Four ecosystem processes: Solar energy flow, the nutrient cycle, the water cycle, and the successional cycle.

Geographic Information System (GIS) software: A digital mapping technology (i.e., software program) that allows you to create, organize, layer, and visualize information based on geography. Google Earth Pro is an example of a GIS software.

Georeferenced: The definition of a single or collection of points to a specific location through map coordinates, which essentially lines up your points to the grid of an underlying data frame, in turn allowing you to query and analyze a collection of points with relation to its surroundings.

Gumbo: A clay-based soil that is characteristically highly elastic due to its ability to absorb vast volumes of water.

Keyline/Keyline design: The practice of making a particular pattern of deep vertical rips into the soil with special shanks to allow water, oxygen, and root penetration deeper into the soil, originally pioneered by Percival Alfred Yeoman.

Lead measure: A predictive measurement of success that you have control of and can manipulate in order to achieve a goal.

Needs and yields analysis: An analysis of the necessary inputs, needs, behaviors, and products of an element within a system in order to put it in the right place relative to other elements in that system.

Open data: Data that is freely available for everyone to use and republish, without any restrictions from copyright, patents, or other mechanisms of control.

Order of operations: In the context of the 5SP Process, this is the hierarchical workflow recommended for diagnosis, design, and implementation. The order is: 1. Geography, 2. Climate, 3. Water, 4. Access, 5. Structures, 6. Fencing, 7. Flora, 8. Fauna, 9. Business, 10. Technology, and 11. Soil.

Perennial crops: Crops that persist between growing seasons and do not need to be replanted/re-seeded.

Personal resources/personal resource inventory: Personal resources include one's living, social, experiential, intellectual, spiritual, cultural resources, financial, and material resources. A personal resource inventory lists *all the things* within these eight categories that could be of use or leveraged to accomplish one's goals, and also any negative resources or debts within those resources.

Photosynthetic capacity: A dynamic measure of the maximum rate at which a plant is able to fix carbon during photosynthesis.

Photosynthetic rate: The rate at which a plant can convert solar light to sugars.

Sector analysis: A representation of the predominant flows of energy towards your property. Some sectors include: flood, noise, erosion/deposition, summer/winter sun/wind, and wildlife.

Six Ps of Epistemology: Epistemology is the study of the nature of knowledge, especially with reference to its limits and validity. In this epistemology model, the six Ps are: patterns, paradigm, philosophy, principle, practices, and process.

Stock: A quantitative measure of an element in your system. A dynamic inventory of what you can see, feel, count, or measure at any given time.

Swale: A shallow level channel with gently sloping sides, designed to manage water runoff and increase rainwater infiltration.

Three permaculture ethics: earth care, people care, and future care. Note that, in *Permaculture: A Designers' Manual*, Bill Mollison described these as "care of the earth, care of people, and reinvestment in those ends." We prefer the former articulation, which is used by many modern-day permaculture instructors.

Upward spiral: A series of thoughts/actions that feeds into itself, causing a situation to become progressively better or more desirable.

Waypoint: A set of coordinates that identify a single point in physical space.

Yield: Any useful resource surplus to the needs of the local system available for use, export, or trade.

Zone five: The unmanaged environment located at the periphery of your property, as it is the least likely visited area. Used for foraging, recreation, or left just as is.

Notes

Introduction

1. Mollison, Bill. *Permaculture: A Designers' Manual*. Tagari Publications, 2012, p. ix.
2. Permares. *Greening the Desert*. YouTube, February 3, 2007. youtube/sohI6vnWZmk
3. Association of Professional Geoscientists and Engineers of Alberta, *Engineering and Geoscience Professions Act*, Section 1, p. 8.
4. Jacke, Dave and Eric Toensmeier. *Edible Forest Gardens*. Chelsea Green, 2005, p. 9.
5. Although Bill Mollison and David Holmgren's *Permaculture One* (Transworld Publishers) was published in 1978, *Permaculture: A Designers' Manual*, which is considered by most to be the foundational textbook, was only published in 1988.
6. Mollison, Bill. *Permaculture: A Designers' Manual*, p. 15.
7. Considine, Bob. "Galileo Wouldn't Have Believed In." *Boston Herald American*, 1974, quote p. 18.
8. "David Holmgren: A Matter of Scale." Interview by Naomi Riddle, *Guernica*, November 26, 2016. Accessed July 23, 2020.
9. Mau, Bruce. *The Incomplete Manifesto for Growth*. Massive Change Network. massivechangenetwork.com/bruce-mau-manifesto. Accessed July 26, 2020.

Step 0: Inspect Your Paradigm

1. Richards, Theodore. *Cosmosophia: Cosmology, Mysticism, and the Birth of a New Myth*. Homebound Publications, 2011.
2. Harris, Sam. *The End of Faith: Religion, Terror, and the Future of Reason*. W.W. Norton Company, 2005.
3. King, Martin Luther, Jr. *The Measure of a Man*. 1959.
4. Daniel Simons. *The Monkey Business Illusion*. YouTube, April 28, 2010. youtube/IGQmdoK_ZfY
5. Horoscopes paint a clear example of this. Should you read a random horoscope to a roomful of people, chances are people will think it's their horoscope that you just read out loud.
6. Encyclopedia Britannica. *Physiology*. Accessed July 31, 2020. britannica.com/science/information-theory/Physiology

7. Suddendorf, Thomas and Michael C. Corballis. 2007. "The Evolution of Foresight: What Is Mental Time Travel, and Is It Unique to Humans?" *Behavioral and Brain Sciences*, 30(3): 299–313.
8. Kahneman, Daniel. *Thinking, Fast and Slow*. 2011.
9. Diamond, Jared. *Collapse: How Societies Choose to Fail or Survive*. Penguin Books, 2005.
10. My first introduction to the concept of upward spirals was a video edit of the book, *Seeing Nature: Deliberate Encounters with the Visible World*, by Paul Krafel (Chelsea Green, 1999). I highly recommend that you watch Paul Krafel's TED talk, "Upward Spirals, Lessons from Nature." youtube/gjy5s2_fyXs
11. Mollison, Bill. *Permaculture: A Designers' Manual*. Tagari Publications, 2012, p. 12.
12. Fragile, resilient, and antifragile is terminology popularized by author Nassim Nicholas Taleb in his book *Antifragile: Things That Gain from Disorder* (Random House, 2012).
13. Meadows, Donella. *Thinking in Systems: A Primer*. Chelsea Green, 2008, p. 160.
14. Gray, Louise. "David Attenborough: Humans Are Plague on Earth." *The Telegraph*, January, 22, 2013. Accessed August 1, 2020.
15. Fox, Matthew. "It's Time to Reclaim a Spirituality of Creation." *OMEGA*, June 10, 2016. Accessed August 5, 2020.
16. Jancovici, Jean-Marc. "How Much of a Slave Master Am I?" Jancovici.com, May 1, 2005. Accessed September 1, 2020. jancovici.com/en/energy-transition/energy-and-us/how-much-of-a-slave-master-am-i/
17. Wright, Ronald. *A Short History of Progress*. House of Anansi Press, 2004.
18. Hitchens, Christopher. "The Catastrophist." *The Atlantic*, January/February 2010. Accessed August 11, 2020.
19. Here is one of many, many studies: Ceballos, Gerardo, et al. "Accelerated Modern Human-Induced Species Losses: Entering the Sixth Mass Extinction." *Science Advances*, Vol. 1, no. 5, June 19, 2015.
20. Krauss, L.M. *A Universe from Nothing: Why Is There Something Rather Than Nothing*. Simon & Schuster, 2012.
21. If you were to condense the timeline of the entire universe and everything that has happened into one calendar year, humans would have invented writing 15 seconds ago.
22. *Collected Works of C.G. Jung*, 2nd Revised ed. Princeton University Press, 2014, p. 43.
23. I realize that this may seem out in left field for some readers, but a surprising number of my students believe that this is indeed the case. I myself, am doubtful.
24. Personal communication with Dr. Richard Bazinet, Associate Professor, Department of Nutritional Sciences, University of Toronto, Canada.

Step 1: Clarify Your Vision, Values, and Resources

1. Seneca, L. *Moral Letters to Luilius, Letter LXXI: On the supreme god* (translator unknown).
2. Australian farmer and engineer P.A. Yeomans invented Keyline design.
3. Roland, Ethan and Gregory Landau. *Regenerative Enterprise: Optimizing for Multi-capital Abundance*. regenterprise.com
4. Mollison, Bill. *Permaculture: A Designers' Manual*. Tagari Publications, 1988. p. 9.
5. Wagamese, Richard. *Embers: One Ojibway's Meditations*. Douglas and McIntyre, 2016.
6. Davis, Wade. "Wade Davis on Endangered Cultures." TED Talk, Vimeo, October 19, 2011. Accessed August 7, 2020. vimeo.com/30794385
7. *Permaculture: A Designers' Manual*, p. 1.
8. Aristotle. *Nichomachean Ethics*. Internet Classics Archive. classics.mit.edu
9. Harris, Sam. *The Moral Landscape: How Science Can Determine Human Values*. Free Press, 2011.
10. Rawl, John. *A Theory of Justice*. Belknap Press: An Imprint of Harvard University Press, 1999.
11. Here are a few of the resources influential in changing my thinking about the sentience of plants: Bird, Christopher and Peter Tompkins. *The Secret Life of Plants: A Fascinating Account of the Physical, Emotional and Spiritual Relations Between Plants and Man*. Harper Perennial, 1989; *The Nature of Things*, "Smarty Plants," Season 51, Episode 17; Buhner, Stephen Harrod. *The Secret Teachings of Plants: The Intelligence of the Heart in the Direct Perception of Nature*. Bear and Co., 2004; Wohlleben, Peter. *The Hidden Life of Trees: What They Feel, How They Communicate, Discoveries from a Secret World*. Greystone Books, 2016.
12. This is a Mollisonian principle from the book, *Permaculture: A Designers' Manual*.
13. Berry, Wendell. *The Gift of Good Land: Further Essays Cultural and Agricultural*. Counterpoint, 2009.
14. Essig, Mark. *Lesser Beasts: A Snout-To-Tail History of the Humble Pig*. Basic Books, 2015.
15. *Permaculture: A Designers' Manual*, p. 507.
16. Diamond, Jared. *Collapse: How Societies Choose to Fail or Survive*. Penguin Books, 2005.
17. Rock, David. "SCARF: A Brain-Based Model for Collaborating with and Influencing Others." *NeuroLeadership Journal*, Issue 1, 2008.
18. I recommend the book *Crucial Conversations* if you want to dive deeper into this strategy. Patterson, Kerry, Joseph Granny, Ron McMillan, and Al Switzler. *Crucial Conversations: Tools for Talking When Stakes Are High*. McGraw-Hill Education, 2nd edition, 2011.
19. Falk, Ben. *The Resilient Farm and Homestead*. Chelsea Green, 2013, p. 47.

Step 2: Diagnose Your Resources

1. Holmgren, David. *Permaculture Principles & Pathways Beyond Sustainability*. Holmgren Design Services, 2002.
2. Miller, G.A. "The Magical Number Seven, Plus or Minus Two: Some Limits on Our Capacity for Processing Information," *Psychological Review*, 63(2): 81–97, 1956.
3. Mollison, Bill. *Permaculture: A Designers' Manual*. Tagari Publications, 1988, p. 70.
4. The 2009 report from the EPA, *Estimating 2003 Building-Related Construction And Demolition Materials Amounts*, estimated new residential projects in 2003 contributed approximately 10.2 million tons per year to landfills, and that renovations to existing residential projects contributed 37.5 million tons of material to the landfill; these stats mark a 155% increase from 1998 with regard to just new residential builds, never mind the renovations—bear in mind, these figures are representative of the US alone. That is not to say that things aren't being done about it; check out endeavourcentre.org
5. Many scientists claim that fungi are closer to animals than plants, and hence why I put them here. You are welcome to put fungi wherever you'd like.
6. *Permaculture Principles & Pathways Beyond Sustainability*, pp. 127–154.
7. Harris, Sam. *The Moral Landscape: How Science Can Determine Human Values*. Free Press, 2011, p. 11.
8. If you are interested in this topic, I recommend these two resources: Orion, Tao. *Beyond the War on Invasive Species: A Permaculture Approach to Ecosystem Restoration*. Chelsea Green, 2015; Dr. Zach Bush, zachbushmd.com
9. Taleb, Nassim Nicholas. *The Black Swan: The Impact of the Highly Improbable*. Random House, 2007.

Step 3: Design Your Resources

1. Jacobs, Herbert. *Frank Lloyd Wright: America's Greatest Architect*. Harcourt, Brace & World, 1965, p. 139.
2. Of course, this would be when I met Michelle in an applied thermodynamics university course during our engineering degrees.
3. Leopold, Aldo. *A Sand County Almanac: With Essays on Conservation from Round River*. Ballantine Books, 1986, pp. 224–225.
4. See the Fisheries (Alberta) Act. qp.alberta.ca/documents/Acts/F16.pdf
5. Chapter 3 of Mollison, Bill. *Permaculture: A Designers' Manual*. Tagari Publications, 1988; and David Holmgren's *Permaculture Design Principles* outlined at permacultureprinciples.com are two excellent resources among many others.
6. Newport, Cal. *Deep Work: Rules for Focused Success in a Distracted World*. Grand Central Publishing, 2016.
7. Or at least, what has become normal.
8. *Deep Work*.

9. This is a skill that needs to be practised, much the same as learning a new instrument.
10. Yes, Michelle and I have been teased and received a good number of chuckles over our choice of that name. We got this idea, and this name, from David Kadavy (kadavy.net/blog/posts/prefrontal-mondays). But better yet, can you guess what we call our Fridays?
11. Lancaster, Brad. *Rainwater Harvesting for Drylands and Beyond*, 2nd edition. Rainsource Press, 2013, p. 152.
12. By far, the most positive, influential, and life-changing experience that both Michelle and I have experienced was from a Vipassana Meditation retreat. If you can make the time to do something like this, I highly recommend it. Alternatively, the Waking Up app, by Sam Harris, is available on iOS and Android and provides an excellent daily ten-minute guided meditation.

Step 4: Implement the Right Design
1. This idea is influenced heavily by the Scales of Landscape Permanence concept first introduced by P.A. Yeomans (*Water for Every Farm: A Practical Irrigation Plan for Every Australian Property*, K.G. Murray, Australia,1973). I'm grateful to many other permaculture designers, teachers, and colleagues who have added brilliant insight to Yeoman's original work, including, but not limited to: Darren Doherty, Javan Bernakevitch, Owen Habutzel, and Geoff Lawton.
2. Churchill, Winston. *House of Commons Rebuilding.* October 28, 1943. Accessed August 11, 2020. api.parliament.uk/historic-hansard/commons/1943/oct/28/house-of-commons-rebuilding
3. McChesney, Chris, Sean Covey, and Jim Huling. *The 4 Disciplines of Execution: Achieving Your Wildly Important Goals.* Free Press, 2012, p. 30.
4. If you'd like to dive into the full-scale framework that I lean into here, seek out Chris McChesney, Sean Covey, and Jim Huling. *The 4 Disciplines of Execution: Achieving Your Wildly Important Goals.* Free Press, 2012. Note that the concept of "weakest link" is borrowed and expanded upon from Allan Savory's framework (Savory, Allan and Jody Butterfield. *Holistic Management: A Commonsense Revolution to Restore Our Environment.* Island Press, 2016).
5. Mollison, Bill. "Phases of Abundance." *Permaculture News*, Permaculture Research Institute, February 9, 2009. Accessed September 15, 2020.
6. *Holistic Management.*
7. At the very least, there's supposedly less than six degrees of separation of all known actors to Kevin Bacon.

Step 5: Monitor Your Resources
1. Mollison, Bill. *Permaculture: A Designers' Manual.* Tagari Publications, 1988, pp. 11–12.
2. Manning, R. "The Oil We Eat: Following the Food Chain Back to Iraq." *Harper's Magazine*, February 2004, p. 37.

3. "'We All Are Implicated': Wendell Berry Laments a Disconnection from Community and the Land." Interview by Scott Carlson, *Chronicle of Higher Education*, April 23, 2010. Accessed September 18, 2020.
4. Aeschylus. *The Oresteia*. Translated by Ian Johnston. Richer Resources Publications, 2007.
5. Li, Q., K. Morimoto, A. Nakadai, et al. "Forest Bathing Enhances Human Natural Killer Activity and Expression of Anti-cancer Proteins." *International Journal of Immunopathology and Pharmacology*, 2007, 20 (2 Suppl 2): 3–8.
6. The journal we were gifted was *The Five-Minute Journal* by Intelligent Change Inc.
7. Hemenway, Toby. *Gaia's Garden: A Guide to Home-Scale Permaculture*. Chelsea Green, 2009, p. 82.
8. Savory, Allan and Jody Butterfield. *Holistic Management: A Commonsense Revolution to Restore Our Environment*. Island Press, 2016.
9. Jones, Christine. "Light Farming: Restoring Carbon, Organic Nitrogen and Biodiversity to Agricultural Soils." amazingcarbon.com, 2018.
10. Siebert, Stefan, Felix Portmann, and Petra Doell. "Global Patterns of Cropland Use Intensity." *Remote Sensing*, 2010, 2: 1625–1643.
11. Morris, Glenn. "Sustaining National Water Supplies by Understanding the Dynamic Capacity That Humus Has to Increase Soil Water-holding Capacity." Thesis submitted for Master of Sustainable Agriculture, University of Sydney, July 2004.
12. Jones, C. "Soil Carbon: Can It Save Agriculture's Bacon?" Paper presented at the Agriculture and Greenhouse Emissions Conference, 2010. Accessed March 15, 2020.
13. Lal, Rattan, Ronald Follett, B.A. Stewart, and J. Kimble. "Soil Carbon Sequestration to Mitigate Climate Change and Advance Food Security." *Soil Science*, 2007, 172: 943–956.
14. Savory and Butterfield. *Holistic Management.*
15. There is clear evidence that humans have managed the entire North American continent for thousands of years (*1491*, Charles C. Mann. Vintage Books. 2006).

The Solution to a Sisyphean Task

1. Korn, Larry (curator). *Sowing Seeds in the Desert: Natural Farming, Global Restoration, and Ultimate Food Security*. Chelsea Green, 2013, p. 16.
2. Kopfer, Manfred. *The Philosophy of Recursive Thinking: A Way Out of the Dead End of Postmodernist Nihilism*. BookRix, 2018.
3. Source of quote unknown.
4. Mueller, Pam A. and Daniel M. Oppenheimer. "The Pen Is Mightier Than the Keyboard." *Psychological Science,* 2014, 25(6): 1159–1168. Essentially, the utilization of day-planners and the relatively slow act of writing facilitates the encoding of information, and is then easier to act upon.

Index

Page numbers in *italics* indicate figures and tables.

A
accountability partners, 19
accuracy, 90–91
animals
   livestock management system, 117–119, 122–123
   relationship with, 55
antifragile systems, 29
aquaculture systems, 119, 121–122
Aristotle, 53
assumptions. *See* paradigms
Attenborough, David, 31
attentional bias, 25
autonomy, 61–64

B
bad decisions, 157–159
bad design, 116, 119–125
bazooka approach, 157–158
beliefs. *See* paradigms
Berry, Wendell, 55, 171
birdshot approach, 142–147, 158–159, 167
Black Swans, 93, 98
brain, 128–129
Brundtland Report, 32

C
CAFOs (Confined Animal Feeding Operations), 117–119
carbon, 188–189
case study review, 136
certainty, 61–64

Churchill, Winston, 144
clarify
   in 5SP Process, 12–13, *14*, 16
   common mistakes, 64–67
   personal resource inventory, 45, *46–47*, 47–49
   practices for, 67, 70–73, 205
   purpose of, 44–45
   values, 50–52, 56
   vision, 56–57, *58*, 59–60
   working with partner, 60–64
climax ecosystem, 189–191, *193*
Coen Farm
   aquaculture system, 119, 121–122
   dam, 98, 124–125
   as ecosystem, 193–195
   food forest implementation, 157–159
   improving productivity, 113–114
   livestock management system, 117–119, 122–124
   mastitis as indicator, 176–179
   succession plan, 184–185
   transition to regenerative agriculture, 38, 40
cognitive biases, 25–26
Collapse (Diamond), 26–27
complex systems, 27, 29
confirmation bias, 25
conflict resolution, 60–64

connections, 166
Cree people, 211–214
cycles, 134, *135*

D

daily reflections, 180–181, 197
dairy cows, 176–179
dam, 124–125
David (statue), 14–16
Davis, Wade, 52
decisions: good, bad, ugly, 156, 159, *160*, 161–165
deep work, 128–129, 135–136
degenerative paradigm
    about, 30–31, *31*
    reflections on, 40–41
    spirals and, 27, 33–34
    transition from, 54
    values, 50
dendritic patterns, 79–82
Dené people, 211–214
design
    in 5SP Process, 12–13, *14*, 16
    adaptable and interactive, 127–128
    deep work, 128–129
    for energy flow, 133–134, *135*, 137
    five questions, 129, 133, 137
    four variables of pattern, 116, *116*, 127
    good, bad, ugly, 116–125
    misconceptions, 111
    mistakes with, 112
    monitoring for bad, 176–179
    practices for, 134–137, 206
    technique vs, *116*
    value of, 108–110
destination, focus on, 18
diagnose
    in 5SP Process, 12–13, *14*, 16
    collection of observations, 87–92
    importance of, 78
    outliers, 93, 98
    practices for, 101–102, 205
    SWOT analysis, 92–93, *94–97*
    value of, 107
    See also information; patterns
Diamond, Jared, 26–27, 59
digital mapping software. *See* GIS software; Google Earth Pro
documentation, 201–204, *205–206*, 207
downward spiral, 27, 33–34, 36
duckweed harvesting system, 123–124

E

ecosystems
    climax, 189–191, 193
    definition, 27, 29
    farms as, 193–195
    four processes, 183, 186–191
Eight Forms of Capital, 47
emotional triggers and explosions, 62–64
endowment effect, 25
energy flow, in design, 133–134, *135*, 137
energy slaves, 34
epistemological model, 9–12
eudaimonia, 53, 56
experimentation, 142–147

F

fairness, 61–64
feedback, 174–175
filestreams, 83, 85–87, 103
financial resources, 45
five struggles of permaculture
    common questions, 8
    determining biggest struggle, 7–9
    first struggle, 44–45
    relationship to 5SP Process, 13, *14*

Five-Step Permaculture Process
(5SP Process)
  development of, 12–13, 14, 16
  summary, 205–206
food, nutrient density, 40
food forest implementation,
  157–159
Forest Garden Community
  Supported Agriculture (CSA),
  194
form, 88–92, 89, 116, 119, 121–122
four ecosystems processes, 183,
  186–191, 192, 197
four variables of pattern, 87–92,
  89, 116, 116, 119–125, 127
fragile systems, 29
Fuller, Buckminster, 111

G
GIS (Geographic Information
  System) software, 98, 99–101,
  103, 108–110, 127–128, 196
Goldilocks zone, 128–129, 181
good, bad, ugly
  decisions, 156, 159, 160, 161–165
  design, 116–125
Good Decision Worksheet, 160,
  161–165, 167, 185
good decisions, 159, 161
good design, 116–119, 127
Google Earth Pro (GEP), 91–92,
  98, 99–101, 103, 196
gorilla in the room, 25–26
GPS apps, 100, 103, 196

H
Harris, Sam, 24, 53, 91
Holmgren, David, 5, 83

I
IKEA effect, 26
implement
  in 5SP Process, 12–13, 14, 16
  experimentation, 142–147
  making good decisions, 156,
    159, 160, 161–165
  order of, 143–144
  practices for, 165, 167, 206
  shotgun approach, 141
  threats to, 140–143
  weekly planning, 151, 154, 155,
    156
indicators, 175, 179, 191, 196–197
industrial agriculture
  four ecosystems processes,
    186–191
  livestock management system,
    117–119
information
  documentation, 201–204,
    205–206, 207
  filestreams, 83, 85–87, 103
  memorization of, 76–77, 79
  organization, 81–82, 101–102
  See also GIS software

J
Jung, Carl, 36

L
Landua, Gregory, 47
Lawton, Geoff, 2–3
lead measures, 149–150, 152–154,
  165, 167
The Limits to Growth (Meadows),
  30
livestock
  management systems, 117–119,
    122–123
  relationship with, 55

M
management. *See* monitor
maps. *See* GIS software
mass extinction events, 34–35
mastitis, 176–179
material resources, 45
Mau, Bruce, 18

Meadows, Donella, 30
measurement
   assessment of, 170–171
   precision and accuracy, 90–91
   See also lead measures
memorization of information, 76–77, 79
Michelangelo Buonarroti, 14–16
Mollison, Bill, 5, 134, 157
monitor
   in 5SP Process, 12–13, *14*, 16
   daily reflections, 180–181
   measurement vs, 170–171
   month-end vision and values assessment, 181–183, *182*
   personal resources, 175, 179–183
   practices for, 196–197, *206*
   property resources, 183, 186–191, *192*
   suffering and well-being, 172–175, 176–179
   ten and two rule, 180
The Monkey Business Illusion (video), 25
month-end vision and values assessment, 181–183, *182*, 184–185, 197
moral landscape, 53

N

needs and yields analysis, 129, 133
nutrient cycle, 183, 188–189, *192*

O

objectivity, 90–91
observation
   effect of paradigms, 90–91
   of resources, 87–92
organic gardening, permaculture vs, 6
organization of information, 81–82, 101–102
outcomes, focus on, 112

P

paradigms
   effect of, 17, 27, 33–36, 90–91
   evolution of, 24–25
   patterns of errors in, 25–26
   practices for, 40–41, 205
   in systems of study, 9, 10–11
   vision and values one-pager and, 66
   See also degenerative paradigm; regenerative paradigm; sustainable paradigm
partners
   for accountability, 19
   conflict resolution, 60–64
patterns
   dendritic, 79–82
   four variables of, 87–92, *89*, 116, 119–125, 127
   in systems of study, 9–10
Peace River, 211–214
perfection, 66–67
permaculture
   5SP Process development, 12–13, *14*
   definition, 29
   industrial practices vs, 117–119, 186–191
   origins of, 5
   prescriptive processes, 11–12
   struggles of, 6–9
permaculture ethics, 45
Permaculture One (Mollison, Holmgren), 5
permaculture property, as complex system, 29
permaculture zones, 158
persistence, importance of, 14–16

personal resources
   categories, 46–47
   collection of observations, 87–92
   inventory, 47–49, 70–71
   monitoring, 175, 179–183
   property resources vs, 83
   SWOT analysis, 92–93, 96–97, 102
   weakest link analysis, 152
"Phases of Abundance" (Mollison), 157
philosophy
   importance of, 24
   in systems of study, 9, 10
photosynthesis, 186–187
placement, 88–92, 89, 116, 123–124
planners, 201–204, 205–206, 207
planning, 116
practices
   summary, 205–206
   in systems of study, 9, 11
precision, 90–91
precision agriculture, 170–171
prefrontal cortex, 128–129
preparation, value of, 107
prescriptions, 11–12
principles, in systems of study, 9, 11
problem solving, 166
process
   focus on, 18
   importance of, 14–16
   in systems of study, 9, 11
property resources
   categories, 82
   collection of observations, 87–92
   monitoring, 183, 186–191, 192
   order for implementation, 143–144
   organization, 82–83, 101–102
   SWOT analysis, 92–93, 94–95, 102, 145–147
   weakest link analysis, 153–154

R
regenerative paradigm
   about, 31, 32–33
   development of, 24, 54–55
   spirals and, 27, 30
   transition to, 38, 40
   values, 51–52
relatedness, 61–64
resilient systems, 29
resources
   framework for determining, 45, 47
   importance of clarification, 44, 60
   living differently, 211–214
   See also diagnose; material resources; personal resources; property resources
Rock, David, 61
Roland, Ethan, 47

S
Savory, Allan, 183
scale, 88–92, 89, 116, 124–125
SCARF social needs, 61–64
sector analysis, 133–134
shields, 134, 135
shotgun approach, 141–151, 157–159
sinks, 133–134, 135
Sisyphus, 172–175, 200
six degrees of separation, 166
Six Ps of Epistemology, 9–11
slug approach, 147–151
social needs, 61–64
societies
   choices to avoid collapse, 59–60

factors leading to thriving/
failure, 26–27
soil
  in design implementation, 145
  health, 113–114
  nutrient cycle, 183, 188–189, *192*
  subsoiling practice, 113–114
solar energy flow, 183, 186–187, *192*
sources, 133, 134, *135*
status, 61–64
storages, 134, *135*
strategy, *116*
struggles. *See* five struggles of permaculture
subsoiling, 113–114
succession planning, 184–185
successional cycle, 183, 189–191, *192*, 193–195
suffering, 172–175, 179
sustainable, terminology use, 32
sustainable paradigm
  about, 31–32, *31*
  reflections on, 40–41
  spirals and, 27, 30, 33–34
  transition from, 54–55
  values, 50–51
swales, 113–114, 124–125
SWOT (strengths, weaknesses, opportunities, threats)
  about, 92–93
  personal resources, 96–97
  property resources, 94–95, 145–147
  review of, 167
systems
  definition, 27
  volatility and, 29

T
Taleb, Nassim Nicholas, 29, 93
technique, *116*
templates, 205–206
ten and two rule for monitoring, 180
timing, 88–92, *89*, *116*, 122–123

U
ugly decisions, 156, 159
ugly design, *116*, 117–119
upward spiral, 27, 34–36

V
value judgements, 92–93
values
  clarification of, 44, 50–52, 56, 60
  paradigms and, 68–69
  reflections on, 67, 70
  See also vision and values one-pager
veil of ignorance, 56
vision
  clarification of, 44, 56–57, 59
  paradigms and, 68–69
  reflections on, 67, 70
  value of, 60
vision and values one-pager
  creation of, 71–73
  example, *58*
  indicators, 175, 179, 196
  mistakes when creating, 64–67
  ranking visions, 59–60
  value of, 57
volatility, 29

W
Wagamese, Richard, 51–52
water cycle, 183, 188–189, *192*
water supply, 29–30, 81, 124–125
weakest link analysis, 148, 152–154, 165, 167
weekly commitment, 150–151, 152–154, 165, 167
weekly planning page, 151, 154, *155*, 156, 165, 167

design their properties for healthier land, healthier profits, and the time to enjoy both.

He currently holds a Permaculture Design Certificate from the Permaculture Research Institute, a Holistic Management Certificate from Holistic Management International, and a Red Seal Journeyman Certificate for Carpentry. His commitment to lifelong learning spans into many other domains beyond agriculture, such as traditional culinary practices, appropriate technology, herbal medicine, shamanism, philosophy, and music.

Learn more about Takota at coenfarm.ca.

Rob and Michelle Avis
Rob and Michelle left the oil and gas industry in 2008 and launched Verge Permaculture, now an internationally recognized and award-winning education company. Through Verge, they are leading the next wave of permaculture education and teaching career-changing professionals how to build successful businesses combining technology with earth science and eco-entrepreneurism.

They also own and operate Adaptive Habitat, a unique and leading-edge property design and management firm through which they leverage their 20 years of combined experience in engineering, project management, ecological design, and sustainable technologies. As skilled Professional Engineers, they offer a depth of practical expertise in building science and appropriate technology (solar, wind, combined heat and power), as well as rainwater harvesting, agro-ecology, ecosystem engineering, soil regeneration, and on-site wastewater treatment/septic design.

Since launching Verge, they've helped more than 5,000 students and a growing number of clients design and/or create integrated systems for shelter, energy, water, waste, and food, all while supporting local economy and regenerating the land. The two have earned testimonials from bestselling author Toby Hemenway, renowned farmer Joel Salatin, among others, and leading permaculture designer/trainer Geoff Lawton who calls Verge "one of North America's premier permaculture design and education companies."

In 2019, Rob and Michelle moved to a 65-hectare (160-acre) wooded property in central Alberta, Canada, and they now spend most of their free time putting the 5SP Process into practice while building their own permaculture property.

Learn more about Rob and Michelle at vergepermacuture.ca.

ABOUT NEW SOCIETY PUBLISHERS

New Society Publishers is an activist, solutions-oriented publisher focused on publishing books for a world of change. Our books offer tips, tools, and insights from leading experts in sustainable building, homesteading, climate change, environment, conscientious commerce, renewable energy, and more—positive solutions for troubled times.

We're proud to hold to the highest environmental and social standards of any publisher in North America. When you buy New Society books, you are part of the solution!

- We print all our books in North America, never overseas

- All our books are printed on **100% post-consumer recycled paper**, processed chlorine-free, with low-VOC vegetable-based inks (since 2002)

- Our corporate structure is an innovative employee shareholder agreement, so we're one-third employee-owned (since 2015)

- We're carbon-neutral (since 2006)

- We're certified as a B Corporation (since 2016)

At New Society Publishers, we care deeply about *what* we publish—but also about *how* we do business.

To download our full catalog, please visit newsociety.com/pages/nsp-catalogue

ENVIRONMENTAL BENEFITS STATEMENT

New Society Publishers saved the following resources by printing the pages of this book on chlorine free paper made with 100% post-consumer waste.

TREES	WATER	ENERGY	SOLID WASTE	GREENHOUSE GASES
138 FULLY GROWN	11,000 GALLONS	58 MILLION BTUs	470 POUNDS	59,200 POUNDS

Environmental impact estimates were made using the Environmental Paper Network Paper Calculator 4.0. For more information visit www.papercalculator.org.

well-being
- as goal, 53, 56
- indicators, 179
- paradigms and, 68–69
- suffering vs, 173–175

What do you have? *See* resources
What do you want? *See* vision
What is right? *See* values
What to do?, clarifying possibilities, 44–45, 44, 60
wildly important goal (WIG), 148–149, 152–154, 165, 167

Z

zone five, 191

About the Authors

Takota Coen

Takota is a fourth-generation farmer and co-owner of Coen Farm, a 100-hectare (250-acre) award-winning organic farm in central Alberta that produces nutrient-dense milk-fed pork, grass-fed beef, pasture-raised eggs, and forest-garden fruits. His family's commitment to increasing the biodiversity of their land and the nutrient density of the food they produce since 1988 has garnered international recognition and numerous environmental awards.

In addition to his ongoing farming experiments, Takota regularly engages with famers, ranchers, and acreage owners across Canada and around the world via his public speaking, farm tours, workshops, private consulting, and online workshops. His life mission is to help stewards everywhere to

Rob Avis, Michelle Avis, and Takota Coen